The Technique of Hybrid Breeding and Cultivating Breakthrough Varieties in Sugarcane

Edited by Wu Caiwen, Yang Kun, Zhu Jianrong, Wu Zhuandi

China Agricultural Science and Technology Press

图书在版编目（CIP）数据

甘蔗杂交育种及突破性品种培育技术＝The Technique of Hybrid Breeding and Cultivating Breakthrough Varieties in Sugarcane / 吴才文等著．--北京：中国农业科学技术出版社，2023.4
　ISBN 978-7-5116-6255-2

Ⅰ．①甘…　Ⅱ．①吴…　Ⅲ．①甘蔗－杂交育种②甘蔗－作物育种　Ⅳ．①S566.105

中国国家版本馆CIP数据核字（2023）第063777号

责任编辑　周丽丽
责任校对　李向荣
责任印制　姜义伟　王思文

出 版 者	中国农业科学技术出版社
	北京市中关村南大街12号　　邮编：100081
电　　话	（010）82106638（编辑室）　（010）82109702（发行部）
	（010）82109709（读者服务部）
网　　址	https：// castp.caas.cn
经 销 者	各地新华书店
印 刷 者	北京建宏印刷有限公司
开　　本	170 mm×240 mm　1/16
印　　张	13.75
字　　数	450千字
版　　次	2023年4月第1版　2023年4月第1次印刷
定　　价	80.00元

◆━━━▌版权所有·侵权必究▐━━━◆

The Technique of Hybrid Breeding and Cultivating Breakthrough Varieties in Sugarcane

Authors

Wu Caiwen[1]   Yang Kun[1]   Zhu Jianrong[1]   Wu Zhuandi[1]
Yao Li[1]   Xia Hongming[1]   Qin Wei[1]   Zan Fenggang[1]
Zhao Jun[1]   Liu Jiayong[1]   Zhao Liping[1]   Li Yuxuan[1]
Wu Jinyu[1]   Zhao Peifang[1]   Zhao Yong[1]   Chen Xuekuan[1]
Eid Mohamed Eid Mehareb[2]   Fouz Fotouh Mohamed Abo-Elenen[2]

---

[1] Sugarcane Research Institute, Yunnan Academy of Agricultural Sciences (YAAS); Yunnan Key Laboratory of Sugarcane Genetic Improvement.
[2] Sugar Crops Research Institute (SCRI), Agriculture Research Center (ARC); Egypt.

BRIEF INTRODUCTION

Sugarcane is the main sugar crop. Improving the benefit of sugarcane breeding and continuously providing more and better varieties for production is the key to promote the sustainable, healthy and stable development of the cane sugar industry. Crossbreeding is the most commonly used, the most common and the most effective method of breeding. This book emphasizes the importance of cultivating independent parent systems, the contribution of peer-to-peer hybridization to sugarcane breeding, the significant impact of "POJ" and "Co" parent systems on sugarcane breeding and cane sugar industry in the world, and the progress of peer-to-peer hybridization in Yunnan; This book expounds the contribution of sugarcane flowering hybridization technology, parent selection method, breeding technology to industrial development, and introduces the achievements of sugarcane variety technology, resistance breeding and new sugarcane variety breeding. The blood relationship and composition of backbone parents of sugarcane breeding in the Chinese mainland are analyzed. This book aims to explore the ways to improve the benefit of sugarcane breeding from selecting original *Saccharum* species in basic hybrid, mating combination, hybrid modes, and hybrid breeding technology.

This book is novel, easy to understand, scientific, accurate, practical, and readable. It can be used as a reference for sugarcane research, teaching, production, and training book.

Preface

This book analyzes the wild composition of hundreds of basic germplasm, hybridization, and sugarcane varieties in the world based on previous studies. It found that: (1) Among more than 300 sugarcane varieties bred by crossbreeding in mainland China, 96 varieties are the direct hybrid progenies of four parents, including 'F134' 'CP49-50' 'Co419' and 'Nco310'. All of them are the progenies of 'POJ2878' and 'Co290'. Other varieties' original *Saccharum* species are almost the same as the fundamental blood relationship of 'POJ2878' and 'Co290'. Similar consanguinity will inevitably lead to shortcomings, such as lack of genetic heterogeneity, low viability, poor adaptability, weak ratoonability, and resistance. (2) The results showed the screening of superior original *Saccharum* species for peer-to-peer crossing, the primordial progenitor heterogeneity between parents was significant. The pedigree was an inverted triangle, the hybridization symmetry was good, the blood relationship was clear, and progeny's heterosis was strong. As parents, more varieties were bred, and the progeny produced more elite varieties, including breakthrough worldwide varieties and parents, which made an outstanding contribution to breeding. The varieties have good traits, significant breakthroughs, a large promotion area, and an outstanding contribution to the cane sugar industry, such as 'POJ2878' 'Co290' 'Co419' 'CP49-50' 'CGT11' and 'CYZ81-173'.

Wu caiwen, the first author of this book, has been engaged in sugarcane crossbreeding for more than 30 years. He has worked in Sichuan Sugar Industry Research Institute for 13 years. He was then introduced to YSRI (Sugarcane Research Institute, Yunnan Academy of Agricultural Sciences) as a senior talent and continued to engage in sugarcane crossbreeding for more than 20 years. Most authors of this book have experiences in India, Brazil, Australia, Thailand, France, and the United States for medium and long-term study and training. They have held many first-hand materials of sugarcane crossbreeding. The writing of this book was supported by the earmarked fund for China Agriculture Research System (CARS-170101), Yunnan Sugarcane Germplasm Innovation and New Variety Breeding Team (2019HC013), and the Sugar Industry Science and Technology Mission of Gengma County, Yunnan Province (202104BI090003).

It is true that, due to the limited level of authors, the omissions in the book are inevitable; some views are relatively new, the understanding among peers may not be the same; in the actual analysis also exist a small number of consanguinity of parents is not very good, but offspring many and good varieties of individual cases, as long as the development of the cane sugar industry is beneficial in the field of sugarcane hybrid breeding continue to find unknown, untied unknown, for the truth and explore, for academic and contention is worth it, ask readers not hesitate to correct, to improve the together further to improve the efficiency of sugarcane hybrid breeding to make more enormous contributions.

Editors
August 2022

Contents

1 Development of Sugarcane Hybridization Technology and Its Contribution to Cane Sugar Industry 1

   1.1 The Origin of Sugarcane Hybridization Breeding Techniques 1

   1.2 The Development of Sugarcane Hybridization Breeding Technology 3

   1.3 Development of Sugarcane Variety Selection Technology 7

   1.4 Contribution of Sugarcane Hybrid Breeding to Industrial Development ...12

   References 16

2 The Technology of Flowering, Hybrid, and Breeding in Sugarcane 21

   2.1 The Technology of Flowering and Hybridization in Sugarcane 21

   2.2 Method and Effect of Sugarcane Melting Pot Hybridization 57

   2.3 Sugarcane Selfing and Intra-Crossing and Their Achievement 62

   2.4 The Method of Selecting Offspring and the Factors Influencing the Selection 68

   References 79

3 The Techniques of Creating Independent Parent Systems and Breeding Breakthrough Varieties in Sugarcane 82

   3.1 Types and Characteristics of Parental Innovation 82

   3.2 Contribution of Independent Parents System to Sugarcane Cross Breeding 86

   3.3 Development of Creating New Types of Independent Parent Systems by Peer-to-Peer Hybridization in Yunnan 91

   3.4 Genetic Relationship and Breeding Effect of Sugarcane Backbone Parents in China 96

   References 115

4 Disease Resistance Breeding and Its Effect in Sugarcane 118

    4.1 Important Sugarcane Diseases and Their Harm 118

    4.2 Disease Resistance Breeding and Strategy in Sugarcane 120

    4.3 Effectiveness and Progress of Disease Resistance Breeding 130

    References 145

5 Stress Resistance Breeding and Its Effect in Sugarcane 149

    5.1 Impact on Cane Sugar Industry of Main Natural Disasters in China 149

    5.2 Drought Resistance Breeding 151

    5.3 Cold Resistance Breeding 163

    References 172

6 Variety Techniques and New Varieties in Sugarcane 177

    6.1 Naming of Sugarcane Varieties 177

    6.2 Symbols of Sugarcane Varieties and Their Breeding Organizations 178

    6.3 Introduction on New Sugarcane Varieties 190

    References 209

1 Development of Sugarcane Hybridization Technology and Its Contribution to Cane Sugar Industry

Sugarcane genetic breeding includes breeding by crossing, transgenic breeding, molecular marker-assisted breeding, and mutation breeding. Sugarcane hybrid breeding is a method for generating new sugarcane varieties by creating new variants of hybrids. Sugarcane is an aneuploidy allopolyploid hybrid; the genetic basis is complex, the interspecific hybrid offspring is large, the variation is many, and the chance of producing good clones is more. Sugarcane sexual hybridization breeding is one of the most commonly used and most popular breeding methods globally. More than 98 percent of sugarcane varieties were obtained by sexual hybridization in China (China Agricultural Information Network, 2013). The process of sugarcane hybridization includes: Firstly, collection, creation, and evaluation of hybrid parents; Secondly, selection of hybrid combinations and production of hybrid fuzz; Thirdly, selection techniques, including a selection of asexual generations (clonal selection), evaluation of ecological adaptability and assessment of disease resistance.

1.1 The Origin of Sugarcane Hybridization Breeding Techniques

The sugarcane hybridization breeding began in 1987−1988. The Dutch Sotwedel and the British Harrison and Boyell had found seedlings of natural hybridization in Java, Indonesia, and Barbados. Then, each of the sugarcane-producing countries was almost all improved by sexual hybridization breeding.

The primary method of sugarcane varieties (especially interspecific hybridization is more important) to breed lots of superior varieties laid the foundation for the cause of world sugarcane breeding.

About ten years after the sugarcane seedlings were discovered, the sugarcane breeder Jeswiet pioneered the sugarcane noble breeding method. In the process of noblization, the tropical species (*Saccharum. officinarum*) with large stem, juicy, high yield, high sugar, and weak resistance is called noble cane, and the small stalk, low sugar, low yield, strong resistance (*S. Spontaneum*) is called the wild cane. The first generation of noble cross F_1 was obtained by crossing the noble and wild species, and the second generation noble F_2 was received by the F_1 back cross. In this way, 'POJ2364' (second generation noble) was obtained by backcrossing 'POJ100' with 'Kassoer' (the first generation of noble species), a natural hybrid of Black Cheribon (*S.officinarum*) and *S. Spontaneum*/Glagah. In 1921, Jeswiet used 'POJ2364' to backcross with another *S.officinarum*, 'EK28', and got a third-generation noble hybrids batch. The cytological analysis of F_1, BC_1, and BC_2 showed that the gamete chromosomes of the noble species' parents doubled after the combination of F_1 and BC_1 gametes. The process of nobilization combined the excellent genetic material of *S.officinarum* into the new hybrids, which was an important discovery in sugarcane nobilization breeding. 'POJ2878' was selected from the noblization process to produce the sugarcane hybrids with high yield and sugar content and superior resistance and adaptability. The breeding of 'POJ2878' is the most significant breakthrough in the cross between *S.officinarum* L.and *S.spontaneum* L. It spread throughout all Java from 1927 to 1930, revitalizing the local sugar industry and becoming the most critical parent of sugarcane breeding worldwide.

In the 1920s, 'Co213' 'Co281' 'Co290' produced in Coimbatore, India, have ternary hybrids of *S.officinarum, S.barberi*, and *S.spontaneum*. from Indian, which are collections of excellent characteristics of disease resistance, high yield, and high sugar content of the above three species. These varieties not only become the main varieties of India but also introduced to South Africa, Argentina, the United States, China, and Australia to promote cultivation, become a worldwide variety, showing a wide range of adaptability of ternary hybrids; these varieties have become the world's major sugarcane a famous hybrid parent of breeding field. The United States Hawaii sugarcane breeding field, with *S.officinarum, S.spontaneum., S.barberi, S.robustum,* and *S.Sinense* 5 kinds of hybrid, breeding 'H32-8560' 'H49-5' and other varieties. Countries worldwide have invested a lot in sugarcane hybridization and new variety breeding. Many new sugarcane varieties with high yield, high quality, and strong resistance have been bred, promoting the sugarcane industry's development in various countries.

Sugarcane crossbreeding requires sexual hybridization, while reproduction is asexual. Each stalk in the variety is the same genotype, which is very consistent unless mutated occasionally. At present, most of the new hybrids worldwide come from the hybridization offspring of 3 to 5 *Saccharum* species. The interspecific hybridization and backcrossing are a recombination of similar genotypes. Inbreeding, narrow genetic basis, near affiliation, resulting in half a century sugarcane breeding in cane yield, sugar content, and resistance, has been difficult to have a more significant breakthrough. Deng et al. (1996) used the Variety Resource Information Management System to analyze the origin of the *Saccharum* species and its progeny of more than 1,000 sugarcane clones and the blood relationship of more than 100 innovative cultivars, found at least seven common ancestors; the blood is very close, the use of wild and wild related species is still minimal.

1.2 The Development of Sugarcane Hybridization Breeding Technology

1.2.1 Photoperiod induction technology and application

Sugarcane is an allopolyploid hybrid or aneuploidy plant with a complex genetic background. Its main economic traits are quantitative traits controlled by micropolygenes and are profoundly affected by environmental factors. In particular, environmental factors restrict flowering habits, Generally in the $8°-10°$ N, sugarcane growth to 3 to 5 internode height, through the natural photoperiod induction, from vegetative growth into reproductive development, flower buds began to divide, and then blossom and bear fruit. However, in the subtropical, sugarcane can not naturally flower, so the region of sugarcane hybridization breeding is minimal (Chen et al., 2003). Photoperiodic induction refers to the photoperiod's effect in a given period, and the photoperiod effect is sustained even after an inappropriate photoperiod condition. Since the 1960s, sugarcane flowering induction technology has become fully developed, respectively, in the United States, Australia, Brazil, South Africa and China Taiwan, Guangxi, Hainan, Fujian and Yunnan Kaiyuan, Yunnan Ruili have built light induction room, parents synchronized flowering, to achieve the purpose of different flowering materials pollination. And further, overcome the problem of interspecific hybridization and then began the diversification of sugarcane breeding objectives.

Photoperiod induction is a standard method to overcome obstacles such as flowering and parents' synchronization. China's first photoperiod chamber for artificial control of sugarcane flowering has been built and used in Guangxi Sugarcane Research Institute (Luo, 1986). Lü (1994) and Hong (1993) placed the difficult-to-flowered fruit sugarcane and sugarcane cultivars under 12 h and 25 min photoperiod treatment, with a blue and far-infrared light at 6: 00−8: 00 am, and 4: 00−6: 15 pm, respectively in the photoperiod chamber as a subsidized light to replace the open-air sunlight, sugarcane cultivars were also controlled by the conventional decreasing day length method, from July 1 to December 20, 1991, the results were first Located in the subtropical region of Zhangzhou City, Fujian Province (24°30′ N, 117°39′ E), it induces local elegant fruit cane in Fujian Province-Tonganguo sugarcane flowering (Lu et al., 1993; Lu et al., 1994). The method induces another sugarcane variety, Maan fruit cane (Lü et al., 1994). Tan et al. (1997) effectively and stably induced the treated materials' flowering by the appropriate coordination of artificial light control, temperature, and humidity. The induction rate was stable from 68.1% to 72% from 1994 to 1996, and the induced material was 75. In the past three years, a total of 305 hybrid fuzz, 225 hybrid combinations, and more than 100,000 seeds with germination ability were obtained, which proved the comprehensive supporting technology for inducing sugarcane flowering under natural conditions in Nanning. Fan et al. (1994) reported the photoperiod induction study of Yunnan Kaiyuan in the high altitude region (23°37′ N, altitude 1,050 m) for three years (1991−1993). These 66 sugarcane varieties could not flower under natural conditions; after photoperiod induction, 22 varieties were flowering, 52 hybrid fuzz and a large number of seedlings were obtained from 44 hybrid combinations, which proved that sugarcane photoperiod induced flowering and sexual hybridization could be successful at high latitude and high altitude. Wang et al. (1999) reported the photoperiod-induced sugarcane flowering in the YSRI in 1998. The two initial treatment times on April 1 and June 1 were 12 h 25 min and 12 h 15 min. The fixed-day treatment, from October 1, the photoperiod of 30 s of decreasing daily light to induce flowering of seven difficult-to-flowering parents and seven ordinary parents, respectively, 'CYT57-423' 'Co419' and 'CMT 70-611', Luohancane 4 difficult-to-flowering parents were persuaded to flower successfully, seven commonly used parents all booting or flowering, and early flowering period of 3 to 4 months. In 1999, the same method (Wang et al., 2000) induced nine difficult-to-flowering parents' flowering. As a result, in the high-latitude and high-altitude Kaiyuan areas, *S.officinarum* 'Crystalina' and Vietnamese Battle Cane had produced the flowering, '48Mouna', and 'Badila' booting for the first time. Also, 'CYT57-423' 'Co419' and 'CMT70-611' were induced to flower again, and the initial flowering period was earlier, and the average flowering rate increased. Li et al. (2000) induced flowering successfully for Badila, 'CYT57-423' 'Co419' 'CMT70-611' and introduced *S.officinarum* '57NG155' 'IJ76-315' 'Korpi', except 'B.Cheribon' and 'Crystalina'. Under appropriate treatment conditions, the flowering rate of these varieties reached more than 30%. The initial flowering period was early to the beginning of November to the beginning of December, effectively solving the problem of no flowering or too late flowering under the natural conditions in HSBS. In 2007, Dong et al. (2008) successfully induced sugarcane varieties 'Badila' and other difficult flowering parents in Ruili, a low altitude area in Yunnan Plateau, for the first time through photoperiod induction. The large-scale application of photoperiod induction technology has produced a large number of hard-to-use parents for parental innovation. With improved photoperiod induction technology and facilities and equipment to improve HSBS, RSBB provides the whole country's hybrid fuzz capacity. In the 2012−2013 hybridization season, the number of parents created by HSBS was 300, and that offered by RSBB was 200. The progress of sugarcane hybridization technology has laid a solid foundation for flowering and utilizing precious germplasm resources and breeding breakthrough sugarcane varieties.

1.2.2 The way and effect of sugarcane parental hybrid

Sugarcane is a highly heterozygous allopolyploid crop; biparental crossing, polycross, selfing, and other different hybridization methods can produce a wide range of offspring separation.

Biparental hybridization refers to how a female hybridizes with a male to obtain a hybrid seed. The method is the most effective hybridization in sugarcane hybrid breeding, the most widely used hybridization. Sugarcane varieties 'POJ2878' 'Co419' 'F134' 'CGT11' 'ROC22' 'CYT93-159' 'CYZ05-51' 'CYZ08-1609' etc. bred by biparental crossing. Besides, in sugarcane germplasm innovation, parent crossing is generally used to facilitate the research and analysis of parents and progenies genetic characteristics. The furnace hybridization method is an improved polycross method (Tew et al., 2011), comprising the hybrid method of multi-females with multi-males and one male with multi-females. To accelerate the determination of parental mating affinity, reduce the number of individual tests and shorten the time of examination of mating affinity, many breeding institutes often choose some varieties with sterile pollen or those treated by emasculation as female parents and take several fine varieties with well-developed pollen as male parents, and randomly put them together to carry out the furnace hybridization method of free mixed hybridization means that each female parent can cross with multiple male parents to obtain hybrid seeds. The Hawaiian Sugar Industry Association used the method as a significant hybrid for commercial breed breeding. The polycross was also used by the Philippines, Sudan, Barbados, and Guadeloupe in France, Brazil, China, and other sugarcane breeding institutes to access some varieties and new materials. However, polycross or furnace hybridization did not become the mainstream method for the world's sugarcane breeding because of the shortcomings of the offspring spectrum.

Self-pollination of monoecious plants and different flowers of males and females in the same plants is called selfing. Sugarcane is a monoecious plant, more comfortable to obtain self-crossing fuzz, the world's major sugarcane breeding institutes have adopted it. Barbados using 'POJ2878' selfing bred 'B37172' and other varieties; Guangzhou Cane Sugar Industry Research Institute (GZSRI) using 'CZZ74-141' selfing produced 'CYT82-339' with traits early maturing, high sugar content, strong ratooning ability, drought-resistance (Zhang, 1995), widely planted in Suixi County and Leizhou City (Feng, 1995). Sugarcane Research Institute, Yunnan Academy of Agricultural Sciences (YSRI) has bred more than 20 clones by selfings, such as 'CYZ71-489' and 'CYR99-634'. Selfing is not significant as a direct method of breeding sugarcane varieties, but it can cultivate excellent parents.

1.3 Development of Sugarcane Variety Selection Technology

1.3.1 Individual plant selection and family selection

Sugarcane sexual hybridization breeding is the primary source of sugarcane cultivars. The selection technology of sugarcane hybrid progeny is the critical factor affecting the efficiency of selecting sugarcane varieties. The selection of sugarcane seedlings is a process of breeding from the sexual generation to the asexual breeding. In Stage 1, since each clone is only one cluster, the selection results can be affected by the environment and competition among clones. A high selection rate(10%−30%) is usually necessary (Marcio et al., 2001). Due to the massive number of sugarcane seedlings, maintaining a high selection rate will lead to a large scale of subsequent clone experiments. At present, the world's selection ways of sugarcane varieties have two methods of artificial selection and family selection. Artificial selection is a method of selecting excellent plants according to individual performance in the field. Family selection is based on the family's average performance in the experiment to choose the good family and select the excellent single plant in the remarkable family. Australia is a developed country of the sugarcane industry in the world. Its sugarcane yield and sugar content are at the highest level in the world. It is mainly due to the successful application of sugarcane family selection technology. The number of varieties bred in Australia is large, and the adaptability is strong, accounting for more than 98% of the national sugarcane area. It has been proved by practice that family selection is superior to artificial selection, with strong breeding pertinence and high breeding benefit. The sugarcane artificial selection method requires breeders to have a lot of time and accumulation experience and needs a strong sense of responsibility and dedication. Still, human life is limited, and breeder turnover often leads to the loss of breeding experience. For a long time, sugarcane varieties in China have been directly selected according to field performance. Since sugarcane is a vegetative propagation crop, the main traits are quantitative, and the quantitative characteristics are susceptible to the natural climate, soil condition, fertilization level, and water status. The impact of management and even the control of the casual will also have a more significant effect on the growth of individual sugarcane, so the breeding varieties are difficult to adapt to different environmental conditions, resulting in low efficiency of sugarcane breeding, high breeding cost, poor adaptability of bred varieties, and small promotion area.

With the development of science and technology, computer's popularization, the endless combination of sugarcane breeding and quantitative genetics theory, the selection of sugarcane varieties can eliminate the environment's influence by using software and appropriate analysis methods to calculate the breeding value. The results showed that the sugarcane varieties bred by family selection technology had good stability and wide adaptability and were easily popularized and applied.

1.3.2 The application of sugarcane family selection technology

Hogarth (1971) first confirms the superiority of sugarcane families' selection and is widely used in the world's major sugarcane breeding institutions. Until 2005, the family selection technology has operated in Australia, West India, Brazil, Colombia, Argentina, Indonesia, Cuba, South Africa, Florida, Hawaii, and Louisiana (Stringer et al., 2011). In 2005, Professor Wu Caiwen of YSRI introduced family selection technology from Australia (Wu, 2007) and successfully used family evaluation technology for the first time in China in 2006.

The family evaluation is not only beneficial to the selection of the clones but also to parental assessment. Stringer et al. (1996) demonstrated that the best linear unbiased prediction (BLUPs) obtained parental evaluation was superior to the experience-based parental review. The choice of the parent directly affects breeding efficiency. The principle of selecting the parent in sugarcane breeding is absolute. Still, the different breeders have different priorities in choosing the parents, and the reaction is not consistent with the breeding effect. Grasping famous parents is beneficial to improve breeding efficiency and reduce blindness (Wu, 2002). The main economic traits and the combining effect of parents on the main economic characteristics of the offspring were studied, which helped improve the selection efficiency and breeding efficiency of sugarcane.

Since sugarcane family selection technology was introduced into China, a large number of parents had been evaluated; for example, Wu et al. (2008) evaluated 24 foreign parents, Wu et al. (2009) evaluated 11 Yunnan innovative parents, Xian et al. (2010) evaluated 100 ordinary sugarcane parents, Qin et al. (2012) and Wu et al. (2010) evaluated ROC series parents, Zhu et al. (2011a; 2011b; 2012) evaluated innovative parents with the blood relationship of *S.spontaneum* and the clones of *S.spontaneum,* Zhao (2011) evaluated 34 Yunnan parents, Zhao et al. (2013) analyzed and assessed the general combining ability and the special combining ability of 6 Q-type varieties and some ordinary parents, which provided a lot of references for sugarcane hybrid combination selection. Besides, based on the family evaluation, Wang et al. (2013) and Xu et al. (2012) also studied some sugarcane parents' economic genetic values.

1.3.3 The progress of sugarcane selection technology and breeding scale

Because of sugarcane parents' high heterozygosity, F_1 hybrids are widely separated, and the favorable gene recombination rate is meager. Only by establishing an appropriate breeding scale can we screen excellent varieties. With photoperiod induction technology, sugarcane breeding can obtain much hybrid fuzz to select sugarcane varieties. For example, the Australian Association of Sugarcane, the annual hybrid combination of the amount of 2,000 to 3,000, of which 1,000 kinds of productive hybrid combinations, exploratory hybrid combinations of 1,000 to 2,000 (Wu, 2007), family evaluation of 400 to 500 crosses, including subordinate four sugarcane breeding stations, each of which has 300−400 hybrid combinations (Wang et al., 2009); St.Gaberiel Research Station, University of Louisiana Agricultural Research Center produces about 400,000 hybrid seeds per year, but due to greenhouse and land restrictions, each year only 100,000 seedlings for sowing, the annual amount of hybrid combinations of 300 to 600. Among them, the proportion of productive hybrid crosses is higher, such as L97-128 has been selected with 234 hybrid combinations, exploratory hybrid combinations of the amount of seedling control in 350 seedlings, while the amount of productive hybrid crosses is as high as 700 (Bischoff et al., 2004); Sugarcane breeding station at U.S. Canal Point is about 100,000 seedlings per year (Tai et al., 2003); Thailand's Mitr Phol Sugarcane Research Center has hybrid combinations of 200 to 500 per year (Weerathaworn, 2006); the Fiji Cane Sugar Industry Institute established 818 hybrid combinations in 2009, of which 561 were multi-male combinations (Sing et al., 2009); the number of seedlings in Stage One was about 100,000 seedlings in Mauritius (Santchurn et al., 2001); Bangladesh as about 100,000 seedlings (He et al., 2010). In Iran, the number of hybrid combinations was only 10 in 1999, and 90 hybrid combinations were introduced in 2000 (Hamdi et al., 2011).

To cultivate suitable varieties of sugarcane as soon as possible, in the period of artificial selection, the domestic sugarcane breeding organizations generally adopted the means of increasing the number of hybrid combinations and the number of seedlings to promote the development of large-scale varieties, speed up the breeding process of sugarcane varieties. For example, from 2010 to 2011, the number of hybrid combinations finished by the sugarcane breeding units was 1,765, and the number of plans was 1,965. Expanding the number of hybrid combinations helps get more good individuals, but requires a lot of money, human resources, and land resources. Therefore, it is essential to study hybrid parents. After introducing family selection technology, the parents with substantial heritability and combining ability are mainly used through many parents' evaluations. The number of hybrid combinations and seedlings gradually reduced. As the breeding efficiency improves, the share of the bred varieties in production is expanding progressively. For example, YSRI only had 30-50 hybrid combinations before 2000, but it had increased to 948 by 2011. Since the family experiment was carried out in 2006, more than 3,000 families have rapidly been evaluated. After comprehensive evaluation and mastering of the heritability, combining ability, and breeding value of sugarcane parents, large-scale optimization of hybrid combinations has been carried out since 2012. The number of hybrid combinations has optimized to 700 and further optimized to 300 in 2019, although the number of hybrid combinations has declined; however, the parents used are the parents with the high breeding value selected from early family evaluation, and the chosen hybrid combination is the combination with high genetic value. Therefore, the breeding effect is better, and the offspring's performance is better.

1.3.4 The ecological adaptability and selection of sugarcane varieties

Sugarcane yield and sugar content are essential indexes for sugarcane breeding, and they are also the two most affected by gene and environment interaction(Jackson et al., 2007). Generally, the popularized sugarcane varieties are self-bred or locally selected (Chen et al., 2012). After the fuzz was obtained from the chosen parents, the clones productivity performance resulted from the interaction between genes and the environment. Ecological breeding, that is, for the complex sugarcane ecological, under different environmental types of breeding varieties, screening the most suitable for local cultivation of improved varieties of sugarcane varieties breeding methods. For example, 4-6 ecological points have been set up according to Australia's environmental type for sugarcane breeding (Hogarth et al., 1989). Taiwan's total area, China, is only 360 thousand km^2, but the soil type and climate of the sugarcane area are different. Taiwan's sugarcane area is divided into 9 regions according to the soil and climate type in 1971. The sugarcane hybridized in the Wantan Sugarcane Breeding Station, and the seedlings were cultivated and selected in the other eight stations (He, 2000).

1.4 Contribution of Sugarcane Hybrid Breeding to Industrial Development

1.4.1 Breeding and planting area of sugarcane new varieties

Sugarcane is native to tropical and subtropical regions. The distribution area is mainly in the north and south latitudes 10°–30°. However, with the development of hybrid breeding technology, the continuous successful hybridization and backcross utilization of wild resources, continuous improvement of varieties and cultivation techniques, the current range of sugarcane cultivation extends to 38° N (such as Spain) and 33° S (Australia), China has reached 33° N (Hanzhong, Shanxi Province), close to the edge of the northern boundary. China located in the northern hemisphere, sugarcane distribution from south to Hainan Island, north to 33° N in the Hanzhong region of Shaanxi, stretch crossing latitude 15°; east to the east of Taiwan, west to southeastern Tibet's the Brahmaputra, across the longitude of 30°, its distribution wide, rare for other countries. The vertical delivery of sugarcane but also continue to challenge the new high, China's Yunnan Province, Yuanjiang, Baoshan, Kaiyuan, Pu'er, and Binchuan counties, there are many sugarcane distributions in the altitude of 1,400–1,600 m plateau, individual sugarcane and even reach 2,000 m, sugarcane is still well behaved, it can get sure that the application of new sugarcane varieties suitable for high altitude areas have expanded the range of cultivable areas.

1.4.2 The regeneration of sugarcane varieties and sugar industry development

As a kind of asexual propagation crop, sugarcane has continuous degeneration due to the continuous accumulation of pathogens in the stems. Simultaneously, due to the social economy's development, the production demand for varieties is higher. Therefore, it is the foundation for the cane sugar industry's sustainable and stable development to continuously select and breed superior varieties with higher yields and better sugar content through hybrid breeding technology. The effect of sugarcane varieties on the development of cane sugar production is very significant; according to statistics, the contribution rate of sugarcane varieties to science and technology is more than 60%. The world has more than 90 countries (regions) grow sugarcane to produce sugar, 2001 the global acreage of 19.2 million hm², with an annual output of 1.247 billion tons of sugarcane, with a yearly production of 103 million tons of cane sugar, accounting for the world's total sugar output of 136.7 million tons 75.7%, To 2010/2011 crush season, the world's sugar production 172 million tons, 80% of cane sugar. China is the world's third-largest sugar producer (Before the 2013/2014 milling season), sugar production of 8.497 million tons in 2001/2002 milling season, of which 87.9% is cane sugar, to 2010/2011 milling season, sugar production has risen to 10.454 million tons, the proportion of cane sugar has increased to 92.4%; therefore, sugarcane hybrid breeding made every achievement much promoted the development of the sugar industry.

In 1921, the Dutch Jeswiet bred a sugarcane variety 'POJ2878' with a giant stalk, high sugar content, disease resistance, and broad adaptability in Java. By 1929, it had almost replaced all the locally cultivated varieties, accounting for 95% of the total sugarcane area in Java, and became the cultivated varieties and hybrid parents of many countries. In the 1950s and 1960s, 'NCo310', a hybrid fuzz obtained from Sugarcane Breeding Institute in Coimbatore, India, was bred in Natel Sugar Institute, South Africa. It was widely planted in South Africa, India, and China, and the cane yield was up to 100 t/hm² in Taiwan, China, More than 70% the local dominant varieties. It had good ratooning and high sugar content, so sugarcane farmers widely welcomed it. In three years, the planting area accounted for 90% of Taiwan's sugarcane planting area. Later, the variety became the primary parent of sugarcane breeding and bred many excellent varieties in Taiwan and other places. After that, according to different ecological regions, Taiwan Sugar Industry Institute implemented the breeding plan of high sugar and plant type selection based on the above varieties as hybrid parents and bred 'ROC1'[①]– 'ROC24' with compact plant type, straight leaf, dark green leaf color, high sugar, strong stress resistance and good ratooning. Among them, 'ROC1' 'ROC5' 'ROC10' and 'ROC16' had successively become the dominant varieties in Taiwan, many of which have also been widely used in the mainland. All countries continuously improve sugarcane varieties as an essential scientific and technological strategy to develop their sugar industry. Australia, Brazil, the United States, India, Thailand, South Africa, and other countries continue to strengthen breeding efforts, at the expense of enormous investment to carry out sugarcane variety breeding and cultivate many new varieties of Q, SP, CP, Co, K, Na series with disease resistance, stress resistance, robust adaptability, and good Ratooning, to reduce the cost of sugarcane planting and sugar production significantly, and improve market competitiveness (Zhao, 2011). In Guangxi, 228 seedlings were cultured with 'CP49-50' × 'Co419'. After eight years of selection, the early maturing, high sugar, high yield, stable yield, rapid budding, high germination rate, fast growth, early closure, high stalk rate, many stems, drought resistance, cold resistance, waterlogging resilience, wide adaptability, and strong ratoonability of 'CGT11' was approved by Guangxi Zhuang Autonomous Region in 1980. This variety has a genetic basis of *S.officinarum*, *S.spontaneum*, and *S.barberi*, which was produced in 1981 and expanded to 47.69% of the total sugarcane area in 1989 (Peng, 1990). By 1990, the acreage of 'CGT11' was about 58% in the Guangxi sugarcane area. Compared with the old variety of 'F134', the proportion of 'F134' land decreased from 96.18% in 1984 to 26.70% in 1990 (Lao, 1998). It was estimated that in 1989, Guangxi planted Tuitang11 133.3 thousand hm², Guangxi farmers increased by 160 million yuan (at that time, the price of raw

---

① ROC 表示新台糖系列品种，全书同。

cane was only 100 yuan/ton), the profit of Sugarmill increased by 32 million yuan, the national tax revenue by 5.1 million yuan, and the commercial profit by 200,772 million yuan (Li, 1982).

The upgrading of sugarcane varieties and the development of the sugar industry in Yunnan have played a massive role in promoting cane yield; for example, Yunnan Province has carried out four large-scale varieties of updates. Each variety update has significantly increased sugarcane yields. From the late 1940s to the mid-1950s, the primary cultivar of Luhancane and Lucane (strawing cane) was about 37.5 t/hm^2. In the late 1950s, 'F134' and 'Co419' were screened and introduced, and by the middle and late 1960s, the average yield was about 45 t/hm^2. In the late 1960s, the selection and promotion of the cultivation of 'CYZ71-388' 'CYZ65-225' 'CMT70-611' 'Chuantang61-408' ('CCT61-408') and 'CYT59-264' new generation of self-breeding, to the late 80s average yield to 50 t/hm^2; By the 1990s, 'CYZ71-95' 'CYZ71-998' 'CYZ81-173' 'CYZ64-24' 'CYZ 89-7' 'CGT11' 'CMT69-421' and 'Xuan 3' etc. were popularized and planted. The average of Yunnan Sugarcane Area increased to about 53 t/hm^2. In recent years, through the promotion and cultivation of new generation of varieties such as 'ROC22' 'ROC16' 'ROC20' 'CYT86-368' 'CYT93-159' 'CYZ89-151' and 'CYZ94-375', the planting area of Yunnan sugarcane has been stabilized at about 300 thousand hm^2 and average was about 63 t/hm^2, The competitiveness of Yunnan cane industry was further improved (Zhao, 2011).

In recent years, the application of new sugarcane varieties has increased the sugar content from 14.42% to 14.81%, and the sugar recovery has risen from 12.30% to 12.98%. To achieve a multi-clones distribution of sugarcane varieties, to accomplish an early-maturing, mature, late-maturing types 4 : 4 : 2 with a reasonable, so that the Yunnan variety structure at the optimal national level, the province's sugarcane sugar recovery for ten consecutive years ranked first in whole China (Wu et al., 2020). At the same time, most of the annual rainfall in the Yunnan west sugarcane area is about

1,200 mm, and the rain is only 1,000 mm since the severe drought in 2009. The annual rainfall in the sugarcane area of Honghe, Yuxi, and Wenshan is only 800−900 mm; The rain is only 500−600 mm during 2009−2013, and about 80% of Yunnan sugarcane planted on dry slope land with no irrigation condition. The breeding and popularization of drought-resistant varieties 'CYT86-368' 'CYZ03-258' 'CYZ03-194' and 'CYZ05-51' has enhanced the risk of drought resistance in Yunnan's sugar industry.

References

BISCHOFF K P, GRAVOIS K A, 2004. The development of new sugarcane varieties at the LUS Agcenter. Journal American Society Sugar Cane Technologists (24):142-164.

CHEN X K, JACKSON P, SHEN W K, et al., 2012. Genotype × environment interactions in sugarcane between China and Australia. Crop & Pasture Science (63):459-466.

CHEN R K, LIN Y Q, ZHANG M Q, et al., 2003. Modern Sugarcane Breeding Theory and Practice. Beijing:China Agricultural Press (in Chinese).

DENG H H, ZHOU Y H, XU Y N, et al., 1996. Analysis on the Genetic Relationships of the Major Sugarcane Clones in China. Sugarcane and Canesugar (6):1-8.

DONG L H, ZHOU Q M, DUAN H F, et al., 2008. Preliminary report of artificial flowering induction for shy-flowering sugarcane parents in the low-elevation inland area of Yunnan Plateau. Subtropical Agriculture Research (3):169-172.

FAN Y H, CHENG T C, WANG L P, et al., 1994. A Study on Photoperiod Induction Technology of Sugarcane Variety Resource. Sugarcane (3):22-27.

FENG Y X, 1995. The Performance and Promotion Prospects of Yuetang82-339 in the Sugarcane Area of Xuwen County.The Science and Technology of Tropical Crop in Guangxi (4):26-28.

HAMDI H, TAHERKHANI K, ALEMANI M P, et al., 2011. Sugarcane Production and Research Activities in Khuzestan-South west of Iran. Balancing Sugar and Energy Production in Developing Countries:Sustainable Technologies and Marketing Strategies, New Delhi, India:39-42.

HE W Z, 2000. Regional Selection of Sugarcane breeding in Taiwan. Guangxi Cane Sugar, 20 (3):61-64, 47-48.

HE W Z, HE H, TANG Q Z, et al., 2010. Summary on the Inspection of Sugarcane Research and Development in Bangladesh. Sugarcane and Canesugar (1):22-27.

HOGARTH D M, MULLINS R T, 1989. Changes in the BSES plant improvement program. Proceedings of the international society of sugarcane technologists (18):956-961.

HOGARTH D M, 1971. Quantitative inheritance studies in sugarcane II. Correlations and predicted responses to selection. Australian Journal of Agricultural Research (22):103-109.

LAO P S, 1998. Has Guitang 11 Degraded? Guangxi Cane Sugar (4):56-57.

LI Q W, DENG H H, ZHOU Y H, et al., 2000. Recent Study on Flowering Induction and Utilization of New Genetic Resources in Sugarcane at Hainan Breeding Station. Sugarcane Industry (1):1-7.

Li Y Y, 1982. The superiority of Guitang11 in Canesugar Production. Guangxi Agricultural Sciences (7):4-7.

LÜ C B, DAI Y M, LIN Y X, et al., 1994. Continuous Report on the Flowering Induction of Sugarcane by blue and far-red light. Sugarcane (3):28-31.

LUO Y G, 1986. The photoperiodic sugarcane chamber has been built and put into use in China. Sugarcane and Canesugar (7):54.

MARCIO H P B, LUIZ A P, LUIS C I S, 2001. Plot size in sugarcane family selection experiments. Crop breeding and Applied Biotechnology, 1 (3):271-276.

PENG S G, 1990. Production performance and status of Guitang11. Guangxi Agricultural Sciences (2):1-4.

QIN W, WU C W, ZENG Q C, et al., 2012. Ratooning traits in sugarcane progenies of ROC varieties as female parents. Journal of Hunan Agricultural University (Natural Science) (1):1-7.

SANTCHURN S, RAMDOYAL K, RIVET L, et al., 2001. Handing large population of sugarcane genotypes at early stages of selection in Mauritius. AMAS. Food and Agricultural Research Council, Reduit, Mauritius:115-125.

STRINGER J K, COX M C, ATKIN F C, 2011. Family selection improves the efficiency and effectiveness of selecting original seedlings and parents. Sugar Technology, 13(1):36-41.

STRINGER J K, MCRAE T A, COX M C, 1996. Best linear unbiased prediction as a method of estimating breeding value in sugarcane. In:Wilson J R, Hogarth DM, Campbell J A, et al., Sugarcane:Research towards efficient and sustainable production. Brisbane:CSIRO Division of Tropical Crops and Pastures:39-41.

TAI P Y P, JR SHINE J M, MILLER J D, et al., 2003. Estimating the family performance of sugarcane crosses using small progeny test. Journal American Society of Sugarcane Technologists (28):61-70.

TAN Y M, ZHANG G M, LIANG L Q, 1997. Study on Synthetic Measure of Flowering Induction and Hybridization of Sugar Cane. Sugarcane (3):1-6.

TEW T L, WU K K, SCHNELL R J, et al., 2011. Comparison of biparental and melting pot methods of crossing sugarcane in Hawaii. Sugar Technology, 12 (2):139-144.

WANG L P, FAN Y H, MA L, et al., 1999. The Research on Photoperiod Induction Flowering of Sugarcane and Their Utilization in Hybridization. Sugarcane (3):1-5.

WANG L P, FAN Y H, MA L, et al., 2000. Photoperiod inducement flowering of shy-flowering sugarcane parents. Sugarcane (3):14-19.

WANG L W, HE H, TAN Y M, et al., 2009. Report on the scientific and technological investigation of Sugarcane and Canesugar in Australia. Sugar Crops of China (2):75-80.

WANG Q N, LIU S M, FU C, et al., 2013. Analysis on Economic Breeding Values of

Usual Sugarcane Parents and Their Crosses. Journal of Tropical and Subtropical Botany (2):155-160.

WEERATHAWORN P, 2006. Research and development work of Mitr Phol sugarcane research center, Thailand. 2006. Proc. Intern. Symp. on Technologies to Improve Sugar Productivity in Developing Countries, Guilin, P. R. China:45-46.

WU C W, 2002. Analysis of the Utilization Efficiency of the Parents for Sugarcane Sexual Hybridization in Yunnan. Sugarcane and Canesugar (4):1-5.

WU C W, 2007. The Technique of Sugarcane Family Selection of BSES in Australia. Sugarcane and Canesugar (1):6-9.

WU C W, LIU J Y, ZHAO J, et al., 2008. Research on breeding potential and variety improvement of exotic parents in sugarcane. Southwest China Journal of Agricultural Sciences (6):1671-1675.

WU C W, LIU S M, LU W X, et al., 2020. The innovation of sugarcane breeding technology and the application of new varieties selection. China Science and Technology Achievements (11):54-55.

WU C W, WANG Y Y, XIA H M, et al., 2009. Research on Heritability and Combing Ability of Creation Parents in Yunnan Sugarcane. Southwest China Journal of Agricultural Sciences (5):1274-1278.

WU C W, ZHAO P F, XIA H M, et al., 2014. Modern cross breeding and selection techniques in sugarcane. BeiJing:Science Press (in Chinese).

WU C W, ZHAO J, ZHAO P F, et al., 2010. Research on Breeding Potential of Several ROC Varieties in Sugarcane. Southwest China Journal of Agricultural Sciences (5):1413-1417.

XIAN W, YANG R Z, ZHOU H, et al., 2010. Study and application of combining ability for sugarcane family selection. Subtropical Agriculture Research (1):4-9.

XU L N, DENG Z H, LIN Y Q, et al., 2012. Series Studies on Economic Genetic Value in Sugarcane (II) Parent Breeding Value and Family Genetic Value Analysis of Sugarcane Seedling.Sugar Crops of China (4):5-9.

ZHANG Y X, 1995. Performance and Cultivation Techniques of Yuetang 82-339. Sugarcane and Canesugar (2):16-18.

ZHAO P F, XIA H M, LIU J Y, et al., 2013. A study of heritability and combining ability of six Q sugarcane varieties. Journal of Hunan Agricultural University (Natural Science) (4):348-353.

ZHAO P F, 2011. Evaluation of the main economic traits and combining the ability of 34 varieties of Yunzhe varieties in Yunnan. Master's Degree Thesis at the Chinese Academy of Agricultural Sciences.

ZHU J R, TAO L A, DONG LH, et al., 2011a. Breeding Potential of *Saccharum spontaneum* L. Collected from Several Ecological Areas of Yunnan. Journal of Southern Agriculture (9):1035-1040.

ZHU J R, TAO L A, DONG L H, et al., 2011b. Breeding Potential of Creation Parents Derived from China Native *Saccharum Spontaneum* in Sugarcane. Journal of Yunnan Agricultural University (Natural Science) (1):12-19.

ZHU J R, TAO L A, DONG L H, et al., 2012. Analysis of combining ability and heritability of *Saccharum spontaneum* L. collected from low and middle altitude areas of Yunnan Province. Journal of Southern Agriculture (3):272-276.

2 The Technology of Flowering, Hybrid, and Breeding in Sugarcane

Sugarcane hybrid breeding is the most commonly used, the most common, and the most effective method globally. Different sugarcane parents cross to obtain fuzz and then selected from the hybrids' offspring to produce new sugarcane varieties that meet the production requirements. Nowadays, almost all Sugarcane varieties planted in growing areas are cultivated by hybridization breeding methods. Since the founding of new China, 315 sugarcane cultivars were planted by sexual hybridization in 320 sugarcane cultivars planted in mainland China, 98% of the varieties, only five varieties bred by radiation and biochemical breeding. Hybridization, selection, and identification, hybrid breeding can get many new functional traits combined with parents and the super-pro-isolated hybrids' genes, especially those related to economic characteristics such as yield, sugar content, height, and diameter. The isolation and accumulation of the genes in the hybrids' offspring may also lead to characters that transcend their parents or produce genetic traits that are not available by genetic interactions. However, hybridization is only a means of promoting the combination of parental genes. Because of the separation and recombination of heterozygous genes, breeders must choose recombination types that meet breeding objectives in this process. Then through a series of tests, selection in the production of the promotion of new varieties, breeding goods traits of continuous breakthroughs and promoting the cane sugar industry's sustainable development.

2.1 The Technology of Flowering and Hybridization in Sugarcane

2.1.1 Flowering behavior of sugarcane

The flowering of sugarcane parents is an essential basis for sugarcane sexual hybridization breeding. The flowering characteristics, heading rate, pollen development, and flowering period of Sugarcane vary significantly among varieties dominated by the species and affected by environmental factors. The flowering pattern of sugarcane in various areas is that the sugarcane has more flowering in the low latitudes than high latitudes. Sugarcane blooms most easily in latitude 10°–20°, followed by 0°–10° and finally 20°–30° (above 20° it decreases with the increase of latitude). In the same area, Sugarcane blooms more in the year if more rainfall in Autumn. According to the different flowering times, it can divide into three types: early flower (November to December), middle flower (December to January of the following year), and late flower (January to February). Because of different varieties, there are also differences in heading and flowering. In general, the clones with the higher consanguineous proportion of wild species have more heading flowers and earlier flowering.

Compared, those with the higher consanguineous percentage of tropical species have fewer heading flowers and later flowering. Some clones bloom only near the equator but not at higher latitudes. Some clones have never found flowers.

After a certain period of growth and sugar accumulation, the growth cones changed qualitatively under suitable conditions, stopped vegetative growth, and turned to reproductive organs' development until heading and flowering. The plant's physiological metabolism also has a series of changes, and the morphology of its growth point will also have corresponding changes. The process of flower bud differentiation, booting, heading, flowering, and fruiting is physiological maturity. Sugarcane from vegetative growth to reproductive growth this process is called the physiological maturity process. The physiological maturity process of sugarcane includes three stages: primordial differentiation stage of young panicle, flowering stage of heading, and caryopsis maturity stage. The differentiation of spikelet primordium can be divided into seven periods: induction stage of flower bud differentiation (IND), initiation stage of inflorescence-axis primordial (IAP), initiation period of inflorescence branch primordial (IBP), initiation period of spikelet primordial (ISP), elongation (ELP), lag period (LG), and period of internode elongation and panicle emergence (IEPE).

From the perspective of external factors, sugarcane flowering has a high demand on the environment. Firstly, sugarcane flowering is sensitive to light. Flowering needs to pass through a certain period of sunshine length, and the flowering will not occur if the sunshine is lower than or higher than a certain length of the light cycle or without a certain length of light cycle change. Secondly, the temperature requirement of sugarcane flowering is also stringent; for example, the temperature in the daytime is higher than 31°C, or the temperature in the night is lower than 18.5°C, which can not be flowering. Therefore, to improve the efficiency of hybrid breeding and use parents effectively, it is necessary to understand the flowering habits and the factors affecting the flowering of sugarcane, create conditions conducive to the flowering of sugarcane so that the parents in need can blossom, the pollen develops nicely. The flowering synchronization serves the hybrid breeding of sugarcane better.

The length of the heading stage of sugarcane varies with varieties, usually 1 to 2 weeks, such as 'F108' generally only seven days, 'S17' is sixteen days, 'F46' even up to thirty-two days, and most varieties of nine to ten days. The heading period also varies with different regions. The flowering period is generally in September near the northern hemisphere equator, 10°N in October, and 20° N in November. The heading occurs in March near the equator in the southern hemisphere and in May to June at 20° S. However, some years of the same variety have heading, and some years have no heading. In natural conditions, most of them do not lead. In Hainan, Sugarcane began to ear at the end of October, the highest in November to December, and the latest in February. The heading rate varies with varieties, such as 'Co290' 'NCo310' and 'ROC10', and the flowering ratio is more than 90%. 'Guangdong10' ('CYT59-65') is the lowest, and the flowering rate is less than 1%. Whether the varieties can ear or not often varies from region to region. For example, 'POJ2727' has few ears in the Philippines, no ears in Florida, and free ears in Coimbatore, India.

Spikelets' flowering order in sugarcane inflorescence is generally from top to bottom and from outside to inside. According to the order of flowering in the heading stage, sugarcane can divide into five types. (1) As soon as the ear is drawn out, it will blossom, such as *S.spontaneum* from India, 'PT28-70', etc.; (2) Flowering occurs at 1/3 of heading, such as 'Co419', 'F134', etc.; (3) Flowering at 1/2 heading, such as 'Co331', 'Co290', etc.; (4) Flowering at 2/3 of heading, such as 'POJ2878', POJ3016', etc.; (5) Not flowering until the heading is over, such as 'F108', 'Co997' and so on. According to the research, the number of spikelets flowering in sugarcane inflorescence is less at the beginning of the heading stage (1) and the end of the heading stage (5) and more at stage (2), stage (3), and stage (4). In a day, different varieties bloom in different periods, 'F108' is the largest in the range of 8: 00 to 13: 00, of which 9: 00 to 10: 00 is the most concentrated, and the rest of the time is only sporadic; 'F134' and 'Co419' have the most flowering at 4: 00−5: 00 daily, while Co351 had the highest flowering daytime from 4: 00 to 7: 00. The most concentrated flowering daytime was from 5: 00 to 6: 00.

Sometimes, sugarcane flowering does not necessarily include pollen, some even if pollen does not significantly develop. Therefore, the technician must check the pollen amount and pollen viability rate before hybridization. There are two ways to check pollen as follows.

Check the amount of pollen: Flowering cane back into the room in the afternoon, water maintenance, put cardboard under the inflorescence in next morning, then gently vibrate the inflorescence, and last classified into five grades: zero, few, medium, many, and much many. Check the pollen development ratio: Use 1g of potassium iodide and 1g of iodine tablet, add 100ml of distilled water to prepare iodine solution. Put the collected pollen on the glass tablet, add a drop of iodine solution, and observe under the microscope. The pollen with spherical shape and blue brown color is the developed pollen, and the pollen with irregular or no blue brown color is the undeveloped pollen. Observe three times continuously and calculate the average pollen development rate. Generally, parents with more pollen and more than 30% development rate can be used as a male parent or female parent, but usually as a male parent. Parents with less pollen and less than 5% development rate are usually used as female parents. The parents used as female parents should be treated to kill all pollens before crossing. The methods of killing pollens include manual emasculation, ethanol emasculation, and warm soup emasculation.

2.1.2 The factors that affect the flowering of sugarcane

The factors that affect the flowering of sugarcane are multifaceted. Due to the different climatic conditions in different areas, sugarcane flowering is different, and Sugarcane is difficult to flowering or rarely flowering in subtropical regions. In the same region, different varieties bloom differently. The identical clones bloom more in the year, with more rainfall in Autumn than in less rain. In other areas, different latitudes, and varying altitudes, sugarcane flowering is also very different. Generally, it is easy to bloom at 10°–20° N and S, next is near the equator, last is 20°–30° area. Therefore, sugarcane's flowering is related to the variety and the climate, determined by internal and external factors.

2.1.2.1 Internal factors

The main factors affecting the flowering of sugarcane are varieties, plant age, leaves, et al.

(1) Variety

There are significant differences in the flowering situation and time of different varieties, mainly determined by the variety. In general, it is difficult for *S.officinarum* to bloom; in addition to 'Uba', it is difficult for 'zhucane' and 'Lucane' to bloom in *S.Sinense*. In *S.barberi*, the Chunnee and Kansar are easy to bloom, but the other clones are difficult to bloom due to poor pollen development and low fertility. In general, *S.robustum* is easier to flower. The *S.spontaneum* is the easiest to flowering. The other related plants, such as *S.Arundinaceum*, *E.fulvus* Nees, and *M.japonica* Ander, are also easy to bloom. Because of the significant difference of flowering difficulty among different species in the sugarcane genus, the flowering habits of varieties bred by distant crossing different species are greatly influenced by their parents. The types with many consanguineous of *S.officinarum* L. are 'POJ3016' 'CYT57-423' 'CYT59-65' 'CYT86-368' 'Q70 B4636' etc., which are difficult to bloom. The varieties with more wild blood relatives such as 'Co33' 'Co421' 'CYR99-113' 'CYR99-601' etc., are more accessible to blossom.

(2) Physiological plant age

The flower bud differentiation of sugarcane is closely related to plant age. The main reason is that sugarcane can feel induced light and produce flower bud differentiation only after reaching a certain plant age. It is believed that the majority of varieties must have 3 or 4 elongation internodes to get this physiological plant age in general. Still, there are differences in this physiological plant age among different types, *S.spontaneum* is the youngest, *S.robustum* is the second, and *S.officinarum* and hybrids need a more substantial plant age. If sugarcane does not reach a certain physiological plant age, the flower buds will not be differentiated even if given the appropriate light and other conditions. The different physiological plant ages in the same plant make each stalk's flowering situation different. Usually, the primary stalk blooms first, followed by the first tiller, the second tiller, etc. The flowering situation is different in different planting periods, the early planting flowers preferably, the late planting flowers later or the following year; the Autumn planting flowers earlier than the spring planting, and the ratooning crop flowers earlier than the plant crop.

The physiological plant age is also related to the effect of the light-induced mutation. In the same condition, the flower bud differentiation of sugarcane with older plant age is not very strict to the light requirement of inducing transformation. The inducing period is also shorter. Under the same condition, it is easier to blossom than sugarcane with shorter plant age.

(3) The relationship between leaves and plant age

The results showed that the leaves in different positions had a significant influence on the flowering physiology of sugarcane. Zhu (1983) thought that the new leaves were substantial to the flowering of sugarcane; they are the central leaves of sugarcane to produce flowering hormone. For example, 'Nco310', the most important is 1 leaf, and 'PR980', the most important is 0 leaf. Suppose their 1 leaf or 0 leaf is removed during flower bud differentiation. If the flower bud differentiation rate is significantly reduced, the heading ratio is low, and the heading time is also delayed. When 1 leaf is unfolding, it is wrapped up with lead paper, shading for a week, and inhibited flowering. Therefore, it is inevitable that 1 leaf is the central functional leaf to produce a flowering hormone.

The results showed that the same grade tiller stalks' flowering was inhibited by shading, and the flowering of the same tiller stalk without shading was normal. It shows that each stem of the same age's flowering process is independent of each other, and the flowering material does not transfer to each other.

2.1.2.2 External factors

The main factors affecting sugarcane's flow are climate, soil, cultivation conditions, latitude, altitude, etc., the most critical factors for the light, temperature, humidity, nutrients, etc.

(1) Effect of light on flowering of sugarcane

A. Effect of photoperiod on flowering of sugarcane

The photoperiod cycle is one of the necessary conditions for sugarcane flower bud differentiation. The results showed that the response of different varieties to the photoperiod is different. The length of sunshine for flower bud differentiation is 12–12.5 h (dark length 11.5–12 h), the most suitable light length is 12 h 28 min (dark length 11 h 32 min). Therefore, 12–12.5 h is the induced illumination range of sugarcane, so sugarcane is called mid-sunshine crops. The light here refers to sunrise to sunset, excluding the rays of before sunrise and after sunset.

The equatorial region's illumination length is 12 h 6 min to 12 h 8 min, and the variation range is minimal. Although the light can cause the differentiation of sugarcane buds in the whole year, the illumination variation deviates far from the most suitable illumination length (12 h 28 min). The effect of the induced transformer is weak (Figure 2-1). The best light for sugarcane flower bud differentiation was from 10° to 20° in South and North latitudes. In the 10° latitude area, the induced light entered the range on July 22 and left on September 9. The period of effective induced flower bud differentiation was 49 days, and most of the time was high-efficient generated light; most of the time, the length of light was close to 12h 28min, so Sugarcane blossomed a lot. The area with a latitude of 20° began to enter into the range of light-induced variation on August 24 and left on October 1 for a total of 38 days. The light-induced change effect was lower than that of 10° but higher than that of 30°. With the latitude increase, the shorter the effective flower bud induced variation period, the worse the effect of induced variation. That's why Sugarcane blooms most in latitude 10°–20°, but not in high latitude areas. HSBS locate at 18°27′N. The period of flower bud induction is from August 22 to October 1, 40 days in total. Most of the flower bud differentiation is around September 6, and most sugarcane varieties can bloom. RSBS (Ruili sugarcane breeding station, Yunnan, China) locate at 24°01′N. The flowering period is from late September to early October, with only more than ten days. Therefore, many varieties can bloom naturally in Hainan but not in Ruili. Although some varieties can bloom in both places, the heading rate is also very different.

According to the change of light length in the whole year, there should be a flower bud induction period in Spring and Autumn. Under the natural condition, only the flower bud induction period in Autumn can produce bud differentiation, and there is no bud differentiation in Spring. The main reason is that the temperature at night is too low, or the age of sugarcane plants is not enough in Spring. Different sugarcane varieties have additional requirements on the illumination of flower bud induction, especially the condition of *S.spontaneum* L. on the day length is not strict. The flower bud differentiation can generally be carried out under illumination longer than 13 h. In Guangdong, Guangxi, Yunnan, Sichuan, and other provinces (autonomous regions), *S.spontaneum* L. successfully blooms in October, and the induction period of flower bud is 13 h or more than 13 h.

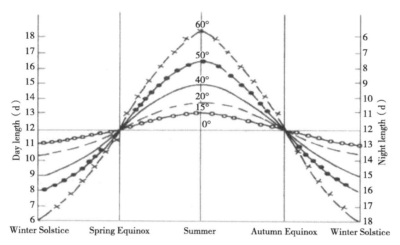

Figure 2-1 Sunshine length curve of north latitude (Wu et al., 2014)

B. Phytochrome

Why does sugarcane flowering require a certain length of illumination? What is the role of induced light? According to modern physiological research, the flowering of plants may relate to a kind of flowering hormone. Under certain conditions, plants can accumulate a certain amount of flowering hormone. When this flowering hormone accumulates a certain amount, the growth point transforms from vegetative organs such as leaves to flower bud differentiation. After several transformation stages of flower bud differentiation, it can be heading and flowering. This flowering-promoting substance belongs to a hormone, so it is called the flowering hormone. It is closely related to the photosensitive pigments commonly found in plants.

Sugarcane, like many higher plants, contains a kind of phytochrome. Its biological characteristics are similar to enzymes. There are mainly two types, namely PR type and PF type, which can transform each other. The maximum absorption spectrum of PR type is 650 nm, primarily absorbing infrared rays. After absorbing 650 nm infrared rays, PR-type photosensitive pigment changes into PF-type photosensitive pigment. The maximum absorption spectrum of PF type is 730 nm, mainly incorporating far infrared. After absorbing 730 nm, far-infrared, PF-type photosensitive pigment changes into PR type. PF type has more vigorous chemical activity and is more unstable than the PR type. It can slowly turn into PR type in the dark, which is called PF type dark conversion. Because the sunlight in the daytime is a diverse spectrum of various wavelengths, including red light and far-red light, the PF type and PR type are constantly converted in the day. Two kinds of photosensitive pigments exist at the same time, but most of them are PF types. Due to the dark conversion of PF type, the photosensitive pigments in plants are mainly PR types. Some people think that PR-type photosensitive pigment is closely related to the formation of flowering hormone in Sugarcane. After a series of physiological changes, PF-type photosensitive pigment can transform into the flowering hormone's precursor. Appropriate PF/PR value forms in the dark transformation process and the flowering hormone is gradually produced (Figure 2-2).

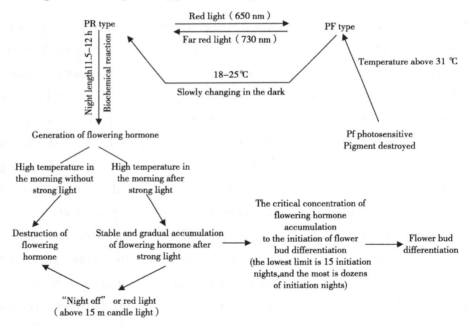

Figure 2-2 Reversible transformation of two types of sugarcane phytochrome and flower bud differentiation (Wang, 1976)

Sugarcane flowering has strict requirements on light length, determined by the time required for the PF type's dark transformation. Because Sugarcane needs a specific dark time, the PF type can slowly transform into a PR type. The experiment shows that the time required for the dark reaction is the same as the dark length (dark period) in the light period (dark period 11.5–12 h). If the time can not reach this dark period, it can not convert into enough PR-type photosensitive pigment, nor can it form flowering hormone. Therefore, the light requirement of sugarcane is the requirement of a particular night length (11.5–12 h). The leaves, mainly the lobus cardiacus and the tender leaves, are responsible for photoinduced transformation and the synthesis of flowering hormones in sugarcane.

The flowering hormone synthesized by leaves must transport to the growing point. The transportation speed of the flowering hormone is slower than that of sucrose and other carbohydrates. According to the test, on the night of flower bud induction, most of the flowering hormones produced by the dark reaction are transported to the growth point before the night of the next day, and the qualitative change occurs at the top when the accumulation of flowering pigment reaches a specific concentration, causing flower bud differentiation. Sugarcane, how many light cycle days (how many inducing nights) are needed to accumulate enough flowering hormone to create flower bud differentiation related to variety, night temperature, inducing light cycle, and light intensity. When the illumination is suitable, the night temperature is high, the daytime illumination intensity is correct, and the variety has no strict requirement on the light period; the night of illumination is less needed; otherwise, it is more. The induced night is less in the low latitude area and more in the high latitude area from the latitude. According to the Hawaiian experiment, it takes 10 to 15 nights (photoperiod) to produce flower bud differentiation, and the period less than ten nights can't bloom.

C. Effect of light intensity on flower differentiation

The photoperiod and light quality (wavelength) can affect flower bud differentiation and light intensity. It has been proved that the distinction of sugarcane flower bud needs two intense light stages. First, sugarcane needs strong light before entering the dark period of induced light because the dark transformation process needs to breathe. Intense light can synthesize more carbohydrates, which provides a sufficient material basis for respiration. Secondly, intense light is also required to transform flowering hormone precursors into the flowering hormone after the dark period. If there is no strong light condition, the dark period's flowering hormone is very unstable. It is easy to decompose under high temperatures. Besides, intense light can promote the transportation of flowering hormones to the growth point. The number of days needed for flower bud induction is closely related to the light intensity. Under the same conditions, when the light intensity is weak in the induction period, a more extended induction period is needed to promote flower bud differentiation; fewer days can be induced when the light intensity is vigorous.

(2) Effect of temperature on flowering and fruiting of sugarcane

The temperature is an essential factor in the bud differentiation of sugarcane. The effect of temperature on the bud differentiation of sugarcane is mainly the temperature at night during the period of bud initiation, that is, the temperature of photopigment transformation from PF type to PR type.

A. Effect of night temperature on flower bud differentiation.

The results showed that the night temperature must be kept above 18 °C for the flower bud differentiation of common sugarcane varieties; once it is below 18 °C, the flower bud transformation cannot be carried out. Therefore, the days with night temperatures below 18.5 °C are closely related to the heading and flowering of sugarcane during the period of flower bud induction. The critical minimum temperature of flower bud differentiation of most sugarcane varieties is about 18 °C, most people thought that the optimal night temperature of flower bud differentiation is 21−26 °C, the critical maximum temperature is about 27 °C, and the night temperature above 27 °C also has an effect on flower bud differentiation. According to the survey data of 11 farms in Maur Island, Hawaii, sugar company, in the northern hemisphere, the low temperature in September is one factor that inhibits sugarcane's flowering. The correlation coefficient between the days when the night temperature is below 18.3 °C and the heading rate is −0.973 from 1st to 21st September (flower bud induction period). In the 21 days, if the night temperature drops to or below 18.3 °C for 11 nights, the sugarcane will not bloom; in 10 nights, it seldom thrives; in 7 nights, it blooms moderately; in 4 nights, it blooms well; in 2 nights, it blooms a lot.

B. Effect of temperature difference on flower bud differentiation.

The investigation shows that temperature difference has a significant influence on the flower bud differentiation of sugarcane varieties. Martin (5° S) in Papua New Guinea and Queensland (17° S) in Australia are where sugarcane is easy to bloom. In sugarcane flower bud induction, the common characteristics are high minimum temperature, slight temperature difference, and more sugarcane flowers (Table 2-1). Simultaneously, the lowest temperature of 102 Field in Hawaii was high. The temperature difference was slight, and the Sugarcane bloomed more; the lowest temperature of Kihei was low, the temperature difference was significant, and the sugarcane did not bloom. The lowest temperature of Port Shepstone in Natal, South Africa, is high. The temperature difference is slight, and the Sugarcane blooms more; the lowest temperature of New Hananes is low, the temperature difference is significant, and the sugarcane does not bloom. The temperature of Pongola is between the two, and the sugarcane blooms moderately (Table 2-2). The results show that the night temperature is too low, the significant temperature difference, and the longer induced change period. Under the ideal night temperature condition, the induction period is shorter. The night temperature, which can not cause flower bud differentiation, is non-inducing night.

Table 2-1 Temperature of flower bud induction in Martin and Queensland (1956–1962)

Unit: °C

Time	Martin, Papua New Guinea			Queensland, Australia		
	Maximum temperature	Minimum temperature	Temperature difference	Maximum temperature	Minimum temperature	Temperature difference
January	30.4	23.7	6.7	31.9	22.5	9.4
February	30.2	23.7	6.5	32.5	22.5	10.0

continued

Time	Martin, Papua New Guinea			Queensland, Australia		
	Maximum temperature	Minimum temperature	Temperature difference	Maximum temperature	Minimum temperature	Temperature difference
March	30.7	23.6	7.1	30.8	21.5	9.3

Source: Peng, 1990.

The effect of night temperature on flower bud differentiation affects the dark period, PF → PR dark transformation. This transformation is a biochemical change process, which is closely related to temperature. Low temperature can inhibit this transformation from preventing flowering hormone precursors and hindering flower bud differentiation.

Table 2-2 Relationship between temperature in the period of flower bud induction and the flowering of sugarcane Unit: °F

Location	Test point	Maximum temperature	Minimum temperature	Temperature difference	Flowering
Hawaii (21.5°N) Mid-August to mid-September	Paia	83.9	70.6	13.3	A lot of
	102 Field	81.9	69.9	12.0	A lot of
	Camp 10	85.4	66.8	18.6	Moderate
	305 Field	82.4	68.0	14.4	Moderate
	400 Field	86.2	65.8	20.4	Less
	Punnene	85.7	69.1	16.6	Less
	Kihei	91.1	65.2	25.9	Not flowering
South African Natal (30° S) March average	Port Shepstone	78.1	67.3	10.8	Many
	Pennington	80.1	67.5	12.6	Many
	Pongola	85.5	66.6	18.9	Moderate
	Mt.Edge Combe	80.6	66.0	14.6	Moderate
	Verulum	83.8	65.7	18.1	Less
	Fshorne	79.2	63.1	16.1	Less to no
	Hill Crest	75.7	59.7	16.0	Not flowering
	New Hananes	81.3	58.8	22.5	Not flowering

Source: Peng, 1990.

C. Effect of temperature on heading, pollen development, and pollination

Low temperature affected the differentiation of flower bud and affected the heading of the flower axis. The results showed that although the flower bud differentiation of sugarcane in Tongshi (high altitude area) was better than that in Yacheng, its heading speed was slower. The reason was that the temperature in Tongshi was lower than that in Yacheng, which resulted in the slow elongation of the flower axis, and it often failed to head before the end of the hybridization season. According to Peng (1990), the pollen development rate of two sugarcane varieties 'Co421' and 'CP29-116', was 24.4% and 10.4%, respectively in the field, and 40.9% and 52.0%, respectively in the greenhouse. The pollen development rate of 'CP49-50' was 58.9%, and another cracking rate of 'CP49-50' was 90.4% at the average temperature of 20.1 °C. Still, at an average temperature of 14.7 °C, the pollen development rate was only 27.7%, and anther cracking ratio was 57.1%. Berding (1983) in Australia Merringa study that 21 °C night temperature can promote sugarcane flowering, improve pollen fertility and seed germination rate, so the evening to 21−23 °C, the day to 27.8−28.9 °C for the appropriate temperature. Low temperature also affects fertilization and seed development, reducing seed germination rate. Therefore, the low temperature during the evening and early morning during pollination in Tongshi, Hainan, is the main limiting factor for the seed setting rate. Practice shows that the seed setting rate of the hybridization in the heated greenhouse is much higher than that in open-air (Table 2-3).

Table 2-3 Effects of temperature on seed setting rate and germination rate of sugarcane seeds

temperature treatment	Pollination period (Dec.24 to Jan. 12)		Mature period (Jan.13 to Mar.6)		Seed setting rate (%)	Germination rate (%)
	Extreme lowest temperature (°C)	The average temperature at 6:00 am (°C)	Extreme lowest temperature (°C)	The average temperature at 6:00 am (°C)		
Open-air	8.9	12.2	3.3	14.2	0.8	5.9
Heating-green house	15.5	17.4	11.7	19.9	4.4	8.5

Source: Peng, 1990.

D. Effects of temperature on stamen development

The impact of low temperature on the stamens is mainly anthers without cracking, pollen grains with little or no development rate being flat, or even not developed. Besides, low temperatures will reduce the rate of pollen tube elongation so that the pollen tube can not reach the ovary, which can not combine with the ovule.

(3) Effects of rainfall and humidity on sugarcane flowering

Rain and atmospheric moisture have a significant impact on sugarcane flowering. Peng (1990) reported Sugarcane varieties such as 'Crystalina' 'Yellow Caledona' 'B. Cheribon' 'Loethers' 'Kassoer' 'POJ3016' 'POJ3303' 'Vietnamese Niucane' 'Guangdong7' 'Guangdong10' 'Co475' and other varieties don't bloom in Yacheng, Hainan, but bloom in Tongshi. Different sugarcane varieties, such as 'Co415' 'Co997' 'F134' 'Guangdong3' 'Guangdong13' 'CYT64-665' and 'CYT62-314' have low and unstable heading rates Yacheng, but high heading rate in Tongshi and bloom every year. The main reason is that the rainfall of Yacheng is less than that of Tongshi, and the relative humidity is lower than that of Tongshi.

The author found that the geographical latitude of Ruili and Kaiyuan in Yunnan is almost the same. Still, the difference in rainfall and air humidity in the two places results in the gap of sugarcane natural flowering performance. Sugarcane flowering is widespread in Ruili's mature production, but Kaiyuan is very difficult to occur (Table 2-4).

Table 2-4 Comparison of rainfall and relative temperature between Hainan and Yunnan during flower bud induction period in sugarcane (September)

Hainan*			Yunnan		
Site	Rainfall (mm)	Relative humidity (%)	Site	Rainfall (mm)	Relative humidity (%)
Yacheng	188.1	85.9	Kaiyuan	90.1	77.3
Tongshi	297.3	88.4	Ruili	166.9	85.3

* Source: Peng, 1990.

Irrigation and watering are good for promoting sugarcane flowering. 'F134' was not flowering in HSBS in 1961, but there are a few flowers in the nearby sugarcane fields with irrigation conditions. From 1962 to 1968, the experiments of irrigation and water spraying were conducted in HSBS and found that moisture conditions significantly affected the flowering of sugarcane. Regarding irrigation time, once a week's irrigation flowering effect is best, followed by once two weeks' irrigation flowering effect is better, and no watering was flowering or very little flowering. Some varieties, such as 'Badila' 'POJ3016' 'Yellow Caledonia' etc., did not promote flowering only irrigation while watering and water spray played a certain role.

(4) Effects of other factors on sugarcane flowering

A. Effects of nitrogen fertilizer on sugarcane flowering

The result of nitrogen fertilizer on the flowering of sugarcane is mainly manifested in two aspects: applying a large amount of nitrogen fertilizer before the transformation period of flower bud reduces or delays the heading of sugarcane. Because nitrogen fertilizer can promote sugarcane leaves to become green, the physiological plant age of the induced light can slow when the plant enters the transformation period. On the contrary, nitrogen deficiency often leads to more heading and flowering.

The experiment showed that the heading rate was inversely proportional to the nitrogen fertilizer applied in HSBS and RSBS. The ratio of heading reduced with more nitrogen fertilizer used. Therefore, to promote the heading of sugarcane, the fertilizer stop period in Yacheng, Hainan, it should not be later than the end of 5 months, and in Ruili, Yunnan, it should not be later than the beginning of 6 months.

B. The relationship between sugarcane heading and the position and slope direction of the sugarcane field

The heading rate of sugarcane in the windward area and windward slope field is generally higher than that in other locations or the center of the sugarcane field. The heading rate of sugarcane in the east, south, and west of the sunward area is higher than in the north. The highest daytime temperature before and during the flower bud differentiation period of sugarcane from August to September is lower, improving the heading rate of sugarcane. With the high altitude, many rainy days, a wet place near the ditch, windward slope, and windward side, the temperature will all be lower, so the heading rate of Sugarcane will be higher.

C. Effects of soil texture on sugarcane flowering

The impact of soil quality on sugarcane flowering is mainly the effect of water and nutrients. Generally, sugarcane blooms more in thin and sandy soil than in fertile soil. According to the investigation of HSBS, Sugarcane in clay soil will sprout later or not; it is easy to bloom in sandy soil. In the garden soil of Coimbatore, India, Sugarcane is more flowering, and the Sugarcane on the moist clay is low or not flowering. The variety of 'D1135' in Puerto Rico had good heading on the humus shale's ridge soil, and the flowering is poor on poorly drained clay. It showed that water in soil affects not only the growth of Sugarcane but also the heading.

2.1.3 The method of regulating sugarcane flowering

Sugarcane flowering in the field of production is unfavorable because flowering consumes nutrients and reduces sugar content. Tall and slender spikes are easy to windbreak and dry to induce lateral branches, resulting in sugar content reduction and cane yield reduction. However, the leading role of sugarcane flowers and seeds is to reproduce offspring. The use of sugarcane flowering and fruiting is the leading way for breeding new varieties. Only flowering can be used as parents for hybridization. Therefore, in the field production, easy flowering varieties to curb flowering and breeding will have to take measures to promote sugarcane flowering, early flowering, or delayed flowering to help flowering synchronization, to facilitate hybridization.

2.1.3.1 Controlling light to regulate flowering

Photoperiod is the most critical condition that causes the differentiation of sugarcane flower buds and can promote sugarcane flowering by light regulation. Most cultivars have an excellent flowering induction effect in the light period of 12 h to 12 h 30 min. In the Canal Point Sugarcane Experimental Station of Florida, the sugarcane was treated with a 12 h 50 min light period from the middle of July. Then the light decreased for 1 min every day for a total of 100 days. In the other treatment, 12 h 30 min was used as the fixed illumination time every day. Many late-flowering sugarcane varieties were induced to bloom early, and the problematic flowering sugarcane varieties could bloom for more than 100 days under 12 h 30 min light. If the artificially induced flower bud differentiation continued to use 12 h 30 min light treatment, it can prevent flowering. From late May to early June, in Louisiana, the United States, sugarcane was treated with 12 h 44 min light, decreasing by 1 min every day, most of the varieties could bloom. In Queensland, Australia, sugarcane was treated with a 1 min decreasing photoperiod per day; Sugarcane has moved into the darkroom at night and out again in the morning, similar to Louisiana's results. In 1964, Sugarcane Research Institute, Guangxi Academy of Agricultural Sciences (GXSRI), used the same method to make 'Co290' 'Co331' 'Co421' 'Nco310' and other varieties bloom.

Wang et al. (1999), a staff member of the Sugarcane Research Institute, Yunnan Academy of Agricultural Sciences (YSRI), used two fixed treatment times of 12 h 25 min and 12 h 15 min as the initial treatment time on April 1 and June 1. The light length decreased by 30 s from October 1 to heading time. Results among the seven difficult flowering parents, four parents were 'CYT57-423' 'CMT70-611' 'Co419', Luohancane bloomed, and seven ordinary parents were all flowering 3–4 months earlier. Cane Sugar Industry Research Institute, Ministry of Light Industry (MSRI) (Li QW et al., 1994) used a fixed time of 12 h 25 min for 'Badila' 'Co419' 'Huanan56-12' ('CHN56-12') 'CHN56-21' 'CYT57-423' and 'CYT59-65' photoperiodic induction treatment, in which 'Badila' had six months of ratoon age from March 1 to September 5 fixed day lengthy processing, September 6 reduced the 30 s per day, the results of booting rate and heading rate of 70%, the average flowering date is December 18, 'CYT59-65' and 'CYT57-423' were treated fixed day long from June 1 to September 5. The daily reduction was 15 s from September

6; the booting and heading effect were the best; the former reached 52% and 48%, respectively. The latter both came to 76%. While 'Co419' was fixed daily length treatments from June 1 to September 5, the booting rate and heading rate reached 100 % after a 30 s reduction from September 6. The best way to treat 'CHN56-12' was the same as 'CYT 59-65', but the treatment effect was better than 'CYT 59-65'; the booting rate and heading rate were 100 %. 'CHN56-21' adopts the same way as 'CYT 59-65' and 'Co419' to induce flowering, with both booting and heading rates reach 100 %.

In the reproductive stage of sugarcane, especially in the early stage of the emergence of mycoplasma spikelets, breaking off the dark with incandescent light at night (called "Breaking Night") can delay the heading with significant effect. In Australia, early flowering varieties can delay flowering effectively by "Breaking Night" treatment during normal flower bud differentiation. Still, some can delay flowering immediately after "Breaking Night" treatment, and others can delay flowering 12 weeks later. There are reports of delayed flowering by "Breaking Night" treatment in reunion the United States and China. Still, due to different varieties, the duration and length of the " Breaking Night " light may be different, which needs to be explored by breeders.

2.1.3.2 Adjusting the planting period to control flowering

To adjust the flowering period and improve the heading rate, the same parent can delay or advance the flowering period through different planting periods. Generally, the flowering period of autumn planting and ratoon crops are earlier than spring planting, and the heading rate is higher. Autumn planting sugarcane is a cultivation system from Autumn to winter (from early August to early November) and the following year. The planting time range of Autumn planting sugarcane is wide. In Autumn, the flowering effect is not the same because of the different planting times and the diverse plant age at photoperiod induction; The harvest time generally starts from November to April of the following year, sometimes postpones to May or June. The previous crop's earlier and later harvest time also influences ratoon crops' earlier and later flowering. Therefore, according to the flowering and hybridization plan, sugarcane flowering time can be achieved through different planting periods in Autumn and harvesting periods of ratoon crops. The results of the experiment in HSBS show that: Sugarcane planted in Autumn had to head 1−26 days earlier than that planted in Spring, and the heading rate increased by 1.7%−43.0% (Table 2-5). From Table 2-6, the heading ratio of sugarcane planted in Autumn is the highest, that of ratoon sugarcane is the second. Sugarcane planted in Spring is the lowest, but the difference among varieties is significant.

Table 2-5 Effects of different planting stages on the heading of variety F108

Year	Planting stage	Beginning of heading	Heading rate (%)
1954	Planted in Spring	Nov. 25	20.1
	Planted in Autumn	Nov. 11	63.1
1955	Planted in Spring	Nov. 17	30.7
	Planted in Autumn	Nov. 13	32.4
1957	Planted in Spring	Nov. 16	29.7
	Planted in Autumn	Nov. 15	23.2
1958	Planted in Spring	Nov. 28	15.2
	Planted in Autumn	Nov. 2	19.4

Source: Peng, 1990.

Table 2-6 Comparison of heading rate of different varieties at different planting stages

Unit: %

Variety	Planted in Spring	Ratoon crop	Planted in Autumn
EK28	32.7	53.7	53.7
F28	26.8	58.0	62.9
F108	15.8	18.0	19.4
Co281	22.4	34.9	64.9
Co290	35.3	36.7	74.9
Co351	20.0	39.6	79.4
POJ234	14.8	30.6	49.5

Source: Peng, 1990.

2.1.3.3 Cut the leaves to adjust the flowering period

Since sugarcane leaves significantly influence flowering physiology, the technician can effectively regulate the flowering period by cutting sugarcane leaves. The results of the 1960's experiment showed that cutting half of the sugarcane leaves can delay flowering for 15–19 days in HSBS. Cut lobus cardiacus and delay flowering for 5–29 days. The results of experiments in Hongguang Farm, Mandan, Yunnan Province, showed that the lobus cardiacus of 'Co285' were cut to delay heading for ten days than that without cutting and cutting lobus cardiacus by two times was half a month later than that without cutting. 'Co312' cut the lobus cardiacus twice and once and then delayed heading for 13 and 14 days, respectively. Zhu (1983) showed that if cutting heart leaves to keep old leaves, flowering is less or not flowering. Sugarcane flowers a lot, if cut the old leaves and keep younger leaves. It is essential to stay-1 blade. If-1 leaf is cut off, the flowering will be severely affected. In Hawaii, rolling up the heart leaf for three weeks from August 15 can prevent flowering. Therefore, cutting blades can delay flowering, but different varieties need to cut different leaves, and when cutting leaves needs to be tested to achieve the desired effect.

2.1.3.4 Promoting flower bud differentiation by increasing night temperature

Generally, from 1: 00–noon, 21.0–28.0 °C is the most suitable temperature for the differentiation of sugarcane flower bud and flowering. In subtropical sugarcane areas, such as the Florida Canal Point in the United States, South Africa, Ruili and Kaiyuan in Yunnan Province in China, due to the high latitude, the night temperature in Autumn and winter is often lower than 21 °C, resulting in male sterility and incomplete pollen development. Therefore, it is necessary to transfer the male parents into the greenhouse to keep warm, prevent male infertility, and improve the pollen development rate. Florida's experience in the United States is that it is better to treat these parents of male sterility or pollen dysplasia with a photoperiod of 12 h 30 min, delaying the flowering for 20–25 days also improve the quality of pollen development.

2.1.3.5 Regulating florescence by water management

Climate or air drying is not conducive to heading, flowering, and seed set. During the initiation period of a sugarcane flower bud (August to September in the northern hemisphere and February to March in the southern hemisphere), it is necessary to ensure proper water supply to promote flower bud differentiation. HSBS test proved that once a week from August to September, water and spray can promote flowering. Yunnan Kaiyuan has low air humidity and low rainfall. It is necessary to spray water as long as it does not rain during flower bud initiation. It is effortless to know that irrigation and spray can promote or regulate the flowering of sugarcane.

2.1.3.6 Promote flowering through nutrition regulation

The practice shows that all the plants with high heading flowering rates are robust. Still, excessive nitrogen use in the high nitrogen condition, especially in the early or middle stage of the induced transformation period, will inhibit sugarcane flowering. The application of nitrogen fertilizer at the seedling stage can promote the plant's growth and the lack of nitrogen fertilizer. Therefore, nitrogen fertilizer can regulate the flowering period by nitrogen fertilizer application. Applying nitrogen fertilizer in the later spikes can promote flowering and seed development and improve the seed setting rate.

2.1.4 Storage technology of sugarcane pollen

2.1.4.1 The significance of research on sugarcane pollen storage technology

The unsynchronized flowering period is a big obstacle for sugarcane hybrid breeding, making many excellent germplasm resources ineffective. On the one hand, there are a lot of germplasm resources. On the other hand, there are some problems such as a narrow genetic basis, single blood relationship, etc. To synchronize the fluorescence, sugarcane breeders use various methods to adjust the fluorescence, and the cost is enormous, such as building a photoperiod room, conserving greenhouse and supporting places, using the Method of increasing or decreasing light to promote or delay the flowering, to achieve the synchronization of florescence. However, this method is still not to help the varieties whose flowering period is too far away (more than two months). The low-temperature and long-term storage technology of sugarcane pollen provide an effective method to solve this problem. Suppose it is successful. The pollen is stored at low temperatures. In that case, It can keep high vitality in months and even years and can hybridize with the female parent at any time, so that the breeder can operate the sugarcane hybridizing work more flexibly, which will bring significant changes to the sugarcane hybridizing work, bring new hope to the effective hybridizing utilization of valuable germplasm resources and the hybridizing combination selected according to the breeder's needs.

2.1.4.2 Research progress of sugarcane pollen storage technology

The research on pollen storage technology of many plants has been successful for a long time. It was reported that the pollens of maize, potato, gladiolus, and plum have higher viability after being stored at low temperature (−196 °C to −40 °C) for 11−24 months. It is predicted that pollen storage time will be prolonged with the development of science and technology, and the pollen survival rate will be improved.

Since the 1930s, sugarcane breeders have made many attempts in sugarcane pollen storage technology. In 1938, Brandes and Sartoris in the United States sent the pollen to keep in a low-temperature container of *S.spontaneum* L. to Columbus by air. These pollens hybridized with 'POJ2725', technicians have produced many seedlings. In 1939, Vijayasaraghy stored sugarcane pollen in a thermos bottle with ice and water, which kept sugarcane pollen's vitality for two days. In 1942, Sartoris reported that after sugarcane bloomed, the pollen stored for 4−6 days could still maintain its viability. In 1962, Coleman used vacuum drying technology to keep sugarcane pollen for seven days at room temperature. After pollination, a small number of seedlings were obtained, but the results were probably self-pollination and cross-pollination. In 1976, Moote found that sugarcane pollen can be stored for 14 days at 100% relative humidity and just a freezing point. However, the above results still fail to solve sugarcane pollen's long-term storage, so it is not good practice for sugarcane breeding. Until the early 1980s, the study had not received substantial progress.

Peng (1990) reported that the sugarcane pollen could survive for 4–6 days according to the research results of G.B. Sartoris, seven days according to R.E. Coleman, and 14 days under the condition of 0 °C according to Moore. In 1983, Dai YB at the U.S. Canal Point used the pollen of 28 *S.spontaneum* clones, 6 *E.arudinceas* clones, 4 *Miscanthus Anderss* clones, and 2 *Sorghum* clones as materials to store. The results showed that the pollen viability of sugarcane and its related plants could be preserved in cold storage at −80 °C for more than 50 days. It was considered that the pollen moisture content was crucial, usually kept below 10%. He H and Gan HP (1990) reported that in 1983, Dr. P.Y.P.Tai of the Canal Point Sugarcane Research Institute of the U.S. Department of Agriculture summarized the previous studies, the technique of long-term low-temperature storage of sugarcane pollen was studied. The results showed that most Sugarcane pollen and the related plants could be stored at −20 °C, −80 °C, and −196 °C for 20–720 days. After hybridization between the stored pollen and commercial varieties, a large number of hybrid seedlings were obtained. This result is a step ahead of the previous studies; breakthroughs have been made.

2.1.4.3 Technical points of sugarcane pollen storage

Sugarcane pollen is highly sensitive to light, temperature, and humidity. Under natural conditions, its vitality can only be maintained for 30 min, then rapidly declining. Therefore, to successfully store sugarcane pollen, one needs to save before and during storage control of pollen vitality factors. To ensure the success of pollen storage, the need to do the following aspects of the work.

(1) Master the time to collect pollen

The anther dehiscence time of different sugarcane varieties, species, and genus is different. Many of them dehisce at sunrise in the morning or sunset in the evening and a few at night. Therefore, it is essential to master the anther cracking time of the varieties (clones) and collect them immediately after anther cracking so that the pollen can enter the subsequent technical treatment before losing its vitality. Tai (1993) thought that one reason why predecessors failed in this study is likely to neglect that pollen can only survive for 30 minutes under natural conditions. Therefore, it is not successful in collecting pollen in time and storing the pollen that has lost its vitality.

(2) Drying treatment of pollen before storage

The water content of fresh pollen varies with the collection environment. Generally, indoor pollen's water content is 4% – 55%, while outdoor pollen is 57%–58%. The survival rate of pollen storage decreases with the increase of the water content of pollen. According to Tai (1993), when the water content of sugarcane pollen exceeds 18 %, the survival rate of storage is meager. Sugarcane pollen must be dried before storage to reduce its moisture content to less than 10 %.

(3) To improve the quality of sugarcane pollen storage, follow the steps below to enhance pollen storage's survival rate

Select the sugarcane plants about to finished heading and blossom, move them into the greenhouse one day before collection, and cut off the sugarcane leaves to avoid affecting the collection operation. In general, an indoor collection environment is better than an outdoor one, which is not easily affected by wind, air humidity, dew, etc., during collection and is free from the influence of external low temperature. He et al., (1990) in the United States Department of Agriculture Canal Point Sugarcane Research Institute test results show that under the same collection time and storage conditions, the indoor and outdoor collection and storage of *S.spontaneum* holes were compared. After 83 days of storage, the pollen was hybridized with commercial variety Co73-378. The number of seedlings per 5 g seed (fuzz) was calculated. The results showed that 233 seedlings were collected indoors, and only 45 seedlings were collected outdoors. The results showed that the effect of the indoor collection was better than that of the outside group.

The plant will be inclined, height and angle are suitable for observation and collection, avoiding pollen loss due to anther opening and plant moving during collection.

Anther dehiscence was observed when most of the pollen dehiscence collect immediately. Usually, use a clean white paper to place under the flower spike, gently shake the flower spike, and make the pollen fall on the paper.

After collection, the pollen shall be sent to the drying room quickly. The drying room shall be kept at a relative humidity of 35 %–55 % and a temperature of about 20 °C. The drying room shall be shaded to avoid the influence of light on the pollen. The anther and other impurities in the pollen shall be cleaned first and then spread thinly to dry. According to the research of Tai (1993), under the above drying conditions, the treatment shall be conducted for 1–4 hours. About 80 % water content of pollen can be removed.

The dried pollen is put into a sealed vial or bag and stored in a refrigerator. To avoid the influence of short-term power failure, it is better to put the small bag or bottle of pollen into a big bottle embedded in ice to prevent the temperature from rising too fast due to power failure and other reasons during storage.

2.1.5 The method of sugarcane hybrid

It is the basis of sugarcane hybrid breeding work to ensure hybridization and improve the quality of hybrid seeds after sugarcane blooms. Sugarcane breeders have done a lot of exploration and accumulated rich experience, making significant contributions to the smooth development of sugarcane hybrid breeding.

2.1.5.1 Intercropping of male and female parents

The adjacent row planting method of male and female parents is also known as the self-rooting crossing method. The specific method is to plant the male parent and female parent adjacent to each other in the field (one row of the male parent, one row of the female parent, cross planting). When the parent is flowering, the flower spike of the parent is close to hybridization. This method can only be used for parents with the same flowering period. If the flowering period is not synchronous, the hybridization is not easy to be successful. In freezing weather during flowering or hybridization, the seed setting rate of hybridization will be significantly reduced. This method is primitive and rarely used at present.

2.1.5.2 Hybrid method of raising stems with sulfurous acid

When the male parent and the female parent bloom, cut down a few feet long panicle stem and insert the male and the female parent into the container containing a 0.03 % sulfurous acid and a 0.01 % phosphoric acid solution time. The panicle can develop normally, and the regular mature seeds can be obtained. For pollination, the male spikes' position should be slightly higher than the female parents' (1/3−1/2 of the male parents' spikes are above the female ones, the same below). To improve the seed setting rate, artificial pollination was carried out in the morning every day. The stem culture solution was added every 2−3 days and replaced once a week. To improve the probability of pollination, the male parent with less pollen can raise several more spikes. After the hybrid pollination, cut the female parent's spikes into bags and dry them in the air when the seeds are mature. If the seeds are not planted temporarily, they should be stored in a paper bag in a desiccator. This method should be carried out in the forest or shade shed to prevent the spikes from dying. This method is convenient for centralized management when many hybrids are carried out; It can improve hybridization efficiency and save labor, which is the most widely used method globally.

2.1.5.3 The method of self rooting of the female parent and sulfurous acid raising of the male parent

This method plants the female parent in the field, and the male parent uses the sulfurous acid solution to raise the stem. The specific way is to select the female parent spike with good growth potential, no pests, and no diseases in the field. Put on a gauze cage 2 to 3 days before flowering to prevent the introduction of other varieties of pollen. When the female parent blooms, select 2 or 3 male parent spikes with good growth to cut down and put the incision into clean water. Use a short-handled sharp knife to cut a slanting mouth again in the water, and then move it into a container filled with 0.01 %–0.03 % sulfurous acid solution to feed the stem of the male parent's spike; then move the male parent's spike to the side of the female parent, and carefully put it into the female parent's cage for hybridization, the male parent's spike should be higher than the female parent's, to facilitate pollination. From 7: 00 to 11: 00 every morning, the male parent's spike can be shaken manually 2 or 3 times to promote pollination. To increase pollination probability, it is better to change a male parent's tassel once every 3–4 days, while the solution for stem cultivation needs to be changed every day. After one month or so of hybridization, the seeds will mature, and then they can be picked and dried to sow and raise seedlings.

This method can improve seed setting rate and quality because the female parents are self-rooting, nutrient supply is good, and pollination can be carried out under normal growth conditions. However, because the hybrid work is carried out in the open field, the workload is large, the hybrid sites are scattered, and the management is not convenient. The hybrid panicles and stems are vulnerable to the windbreak. Simultaneously, the flowering and hybrid sugarcane are vulnerable to low temperature and freezing damage in winter. If the night temperature is lower than 18 °C, pollen viability will be significantly reduced, which will affect the hybridization effect, so it is seldom used at present.

2.1.5.4 Method of high pressure rooting of the female parent's phimosis and sulfite cultivation of male parent's stem

In the parent plot, the plants with prominent booting characteristics are selected as mating stems. The rotten wet compost is wrapped on the 2 to 3 nodes in the middle and upper part of the spike stem with film, and the upper and lower ends are tightly tied to keep moisture and heat to promote the rooting. When more fibrous roots are produced in the phimosis, the lower part of the phimosis is cut off. The cut spike stem is soaked with water. The film is removed and carefully moved to the hybrid shed. It is planted in a jar or plastic bucket. When flowering, it can be hybridized, and the spike stem of the male parent can be raised with the sulfurous acid solution. When hybridizing, the parents should be close to each other and put on a gauze cage to facilitate pollination hybridization and prevent other pollens' spread. After pollination, the seeds mature for about 30 days, and the seeds can be harvested and dried for sowing.

This method is convenient in work, less in workload, easy to manage, saves labor and financial resources, and avoids wind damage. Low temperature freezing damage can also transfer the male parent to the conservatory to reduce the freezing damage. It is a better hybrid method, which is used in many sugarcane breeding organizations.

2.1.5.5 Furnace hybridization method

Mongelsdof created this method in the United States to accelerate the determination of parental mating affinity, reduce individual measurements, and shorten the time of mating affinity testing. Including the following two methods:

(1) The hybrid method of multi-female parents and multi-male parents

Some sterile pollen varieties, or the ones treated by emasculation, were selected as the female parent. Several fine varieties with good pollen development were selected as the male parent. They were randomly put together and allowed to mix and cross freely. As it takes 5-10 days for the flower spike to open, its position frequently changes. The seeds of the female parents can be harvested at the same time. The offspring only know the female parent, but not the father.

(2) One male and many female hybrid methods

The varieties with proper pollen development were selected as male parents, and some were wholly undeveloped or castrated as female parents. After pollination, the male flower spikes were removed, and only the seeds of the female parent were harvested. Therefore, these offspring knew their parents.

2.1.6 Selection principle of sugarcane hybrid parents

The selection of hybrid parents and combinations is the critical link of sugarcane hybridization breeding. The choice of a tasty sugarcane variety has to undergo complex processes such as hybridization, selection, identification, and reproduction. There are many uncertainties in this process, such as the parent's use, the choice of the combination, the cultivation of the offspring, the method of choice, and identification. The practice has proved that selecting hybrid parents and combinations is key to cultivating excellent sugarcane varieties. If the hybrid parents' choice well, the combination of sound, unique opportunities for future generations of variation, the breeder can breed more varieties, and future generations of the more breakthrough, the more energetic.

On the contrary, if the parent selection and the matching combination are incorrect, it is difficult to cultivate a tasty variety, resulting in human, material, and financial waste. Therefore, the valid selection of parents, reasonable selection of combinations is crucial for sugarcane hybrid breeding. The choice of the parents directly affects breeding efficiency. The principle of selecting the parents in sugarcane breeding is individual. However, due to the different breeding objectives, the focus of choosing the parents is different. The reaction is not consistent with the breeding effect.

2.1.6.1 Increase the kinship of parents and select the hybrid combination with high heterogeneity

The heterogeneity of sugarcane parents is mainly derived from different sugarcane genera species, such as *S.officinarum*, *S.barberi*, *S.Sinense*, *S.spontaneum*. and *S.robustum*, and Jeswiet. With the continuous progress of distant interspecific hybridization, the blood relationship of wild germplasm of related genera such as *E.arudinceas*, *Erianthus arundinaceum* J., and *E.rockii Keng* is also steadily infiltrating into sugarcane varieties, and the character of sugarcane varieties is continuous to make the breakthrough. The higher the parents' heterogeneity, the richer the offspring's genetic content, the more the gene recombination, the stronger the heterosis. The cultivation of the breakthrough varieties relies on the discovery of the breakthrough germplasm. From the end of the 19th Century to the beginning of the 20th Century, many sugarcane parents were created and selected through natural pollination, artificial hybridization, and artificial selection. For example, 'POJ100' and 'Ek2' are inbreeding of two *S.officinarum* L; 'Kassoer' is valuable essential germplasm produced by the interspecific hybridization between *S.officinarum* L and *S.spontaneum* L from Java, Indonesia. Due to their lack of blood cross and significant heterogeneity, these basic germplasms have bred many breakthrough sugarcane varieties, such as 'POJ2878' and 'Co290', etc., through reciprocal hybridization. Through continuous hybridization and backcrossing, many breakthrough sugarcane varieties/parents, such as 'Co419' 'NCo310' 'CP49-50' 'F134' and 'CGT11' etc. However, some parents did not breed any varieties when they crossed with each other. For example, the cross between 'CGT11' and 'F134' did not produce any variety. The main problem is that they are all the offspring of full sibs. The same parents were only the exchange position between males and females. Besides, 'Co419' and 'CP49-50' direct hybridization in China also cultivated 12 excellent varieties. Still, their hybrid offspring between the hybrid has not developed any variety, although these parents of good traits, but no heterogeneity, offspring is also challenging to cultivate excellent variety. Therefore, increasing new consanguinity and making full use of the original ancient, which has not been used, will breed better sugarcane varieties.

2.1.6.2 Selecting parents and mating combinations according to breeding objectives

Sugarcane varieties grow for a long time, generally up to 1 year or even more. A variety should be popularized and applied in a sugarcane area. It should not only adapt to the natural climate conditions of all seasons in the sugarcane area but also adapt to all kinds of abiotic natural disasters (such as drought, cold disaster, typhoon, etc.) and biological disasters (such as pests, diseases, weeds, and rats, etc.) in the sugarcane area. Therefore, sugarcane varieties have solid regional characteristics. Different sugarcane regions have diverse requirements and breeding objectives for improved varieties due to different environmental conditions, productivity levels, and farming systems. According to breeding objectives requirements, parents and combinations must be chosen to select new varieties suitable for the local climate and sugar production technology. For example, the dry slope area accounts for 80% of the total area in the Yunnan sugarcane area, the soil is poor, the irrigation conditions are poor, the cultivation is rough, and the management level is low. This sugarcane area must select varieties with excellent resistance to barren, drought, and ethical rationality; attention should also be paid to these characteristics when selecting parents. For example, 'CYZ03-194' 'CYZ03-258' and 'CYZ06-407', which 'ROC25' chose as parents, are not only drought resistant, ratooning and adaptable, but also widely used in dry slope land of sugarcane areas. Simultaneously, there are still 20% sugarcane areas with good water and fertilizer conditions in Yunnan. These areas that have the right ecological and natural conditions are high-yield and high-sugar areas.

2.1.6.3 Selecting combinations with parents with many excellent characters and good character complementarity

The principal contradiction of character complementation should be grasped according to breeding objectives, especially the main characters that limit the further improvement of yield and quality. In general, the complementation of yield components should be considered first. For example, when the structure of yield components required by breeding objectives is the type of effective stem and large stem type. The parents of large stem types and multi-stem types can cross each other. Secondly, we should consider the characters' complementary (such as disease resistance, drought resistance, cold resistance) that affect stable yield and quality characters. When the breeding goal requires a significant character breakthrough, the parents who must be selected have better performance and complementarity.

However, two parents' complementation traits have a specific limit, and one of the parents cannot have too serious disadvantages, especially in the essential characters. It can't have shortcomings that are difficult to overcome. Simultaneously, the complementary traits between parents should not be too much to avoid the severe separation of hybrid offspring, increased separated generations, and the extension of breeding years. If the evil characters have little impact on the local sugarcane production, the breeders can also not consider the effect of the offspring's character when selecting combinations; for example, there is no smut pathogen in Australia. As long as the pathogen is not brought in, the disease will not occur in the sugarcane area. When selecting parents, the resistance of parents to smut may not be considered. No variety in production can be perfect, with both good and evil characters. When a parent lacks a specific trait, another parent combination with this character should be selected; Thus, it is possible to breed varieties with better characteristics.

2.1.6.4 Selection elite production varieties as parents

The adaptability of cultivars to external conditions is an essential factor affecting high yield, stable yield, and quality traits. Whether the hybrid offspring can adapt to the local requirements is very important to the parent's adaptability. In areas where natural conditions are harsh, cold, and drought, local cultivars are often more adaptable than alien varieties due to long-term biological adaptation and artificial selection. In this area, it is better to use popularized varieties as parents. They have specific adaptability to local natural conditions. They can maintain the most excellent characteristics, but it is also conducive to generating traits and cultivating better varieties. Therefore, exploring good parents can effectively improve breeding efficiency from the local popularized varieties and their offspring. For example, among the 32 sugarcane varieties cultivated in Yunnan, 31 are used as parents, 19 of which are widely used in production. Other parents have introduced clones with complementary characters or offspring of wild relatives, which breed better varieties and increase the adaptability and high yield of bred varieties.

2.1.6.5 Selecting clones with many ecological types, significant genetic differences, and far genetic relationships as parents

Varieties with different ecological types, geographical sources, and genetic relationships are used as parents; it is easy to select elite varieties with characteristics beyond parents and strong adaptability because of the significant difference in genetic basis and the vast separation of the hybrid progeny. Sugarcane hybrid breeding practice has proved that under normal circumstances, using different types of ecological clones as a parent is easy to overcome local parents' limitations or shortcomings to increase the chances of success. For example, among 241 sugarcane varieties bred by cross-breeding in China, 200 parents are directly bred using foreign and domestic varieties. The rest 40 varieties contain foreign species. The higher the genetic difference is, the greater the genetic variation will cause, the more significant the segregation of the hybrid progeny characters, and the longer the generation of separation, which will affect the efficiency of breeding. For example, the original clones of the hybrid progeny of sugarcane species have been crossbreeding with E.Arundinaceum for half a century, and no good variety has been bred up to now. 'Yunnong01-58' ('CYN01-58') is a new sugarcane variety, bred by Yunnan Agricultural University used 'CYC89-8' crossed with *E. Arundinaceum* of Kunming. Still, only in high altitude areas, once the altitude is reduced, the production performance will decline instead, and there is no promotion value. In general, when parents are selected for super parent breeding and cultivating breakthrough sugarcane varieties, the higher the genetic gap between parents is, the better is required for extensive interspecific hybridization in Sugarcane. Although geographical distance or geographical gap in early sugarcane breeding sometimes reflects their genetic differences, there is no direct relationship between them. Especially in recent years, the introduction of sugarcane is frequent, and germplasm resources are often shared worldwide. After many varieties have been improved, it isn't easy to judge the distance of their genetic relationship from the geographical position. The genetic difference between parents can't depend on the distance between parents.

2.1.6.6 Parents have good combining ability

Parents' excellent traits and fewer shortcomings are essential for parents' choice, but not all suitable varieties are perfect parents. For example, in the 1960s, 'CGT11', a superb variety bred by GXSRI, was planted in China's leading sugarcane producing areas. Many breeders used it as parents and matched many cross combinations, but few of them succeeded. Sometimes a variety that is not outstanding in itself is a good parent. For example, 'CHN56-63', bred by Guangdong Academy of Agricultural Sciences in China, is not a tasty variety in itself and has not been widely planted in the sugarcane area. However, it is used as a parent to cross with other varieties. Many suitable varieties with high yield, high quality, and solid cold resistance have been bred in China's northern sugarcane region. In recent years, the concepts of combining ability, breeding value, and economic breeding value have been introduced into sugarcane crossbreeding. Considerable progress has been made in selecting excellent parents and selecting combinations by evaluating the combining ability, breeding value, and parents' economic breeding value. For example, R software has been used by YSRI; a group of sugarcane parents with strong ratoon, drought resistance, high yield, and high sugar content were selected by the method of family evaluation. Therefore, sugarcane breeders should pay attention to sugarcane varieties' advantages and disadvantages and choose good parents through crossbreeding practice and modern science and technology to obtain better results.

Besides, the use range of parents should be adjusted according to the development level of productivity. For example, if a sugarcane area wants to improve the sugar content significantly, it is necessary to strengthen high sugar parents' use frequency. With mechanization development, selecting parents with excellent lodging resistance, strong rooting, and defoliation is conducive to cultivating varieties suitable for mechanized production. Due to the difference in pollen development of parents, The paternity of parents should also be adjusted. In general, those with more pollen should be used as males, while those with less pollen should be females.

2.2 Method and Effect of Sugarcane Melting Pot Hybridization

2.2.1 The concept of melting pot hybridization

Method of melting pot crossing (modified polycross) is an improved multi-hybrid method (Tew et al., 2011), which includes multi-female parents with multi-male parents and one-male parent with multi-female parents. In the multi-father-multi-mother hybridization method, some sterile pollen parents or emasculation are used as female parents, and parents with abundant pollen are treated as male parents. They are randomly put together to be freely mixed and hybridized. Because it takes 5–10 days for the flower spike to bloom, the flower spike's position needs to be changed frequently, the seeds of parents can be harvested simultaneously, so only female parent is known for their progenies. In the method of one-father-multiple-mother hybridization, a variety of excellent pollen development is selected as the male parent. Several varieties with no pollen development or several varieties treated by emasculation are used as the female parent. After flowering and pollination, the male parent's tassels are removed. Only the seeds of the female parent are harvested, so parents' information in this type of hybridization. In 1953, Dr. Mongelsd of the Hawaii Sugar Planters' Association first proposed the hybridization of sugarcane. He thought that sugarcane was diploid, and there should be high genetic variation among the progeny populations, so a considerable breeding population (1 million seedlings) should be established (Chen et al., 2003). Melting pot hybridization can easily obtain a large number of seedlings and save a lot of labor costs. For example, from 1942 to 1946, all the Hawaii Sugar Association's fuzzy was produced through melting pot hybridization (Tew et al., 2011). Compared with two-parent hybridization, the cross-computation design could be more predictable according to their parents' traits, genetic basis, flowering period, etc., which has strong pertinence.

Combining emasculation can improve the hybridization effect, but it needs more facilities, equipment, workforce, and material resources. However, the melting pot hybridization method is relatively simple, saves workforce and material resources, and obtains more seedlings.

2.2.2 The effect of melting pot cross breeding

In Hawaii, USA, melting pot hybridization is the most crucial method, and the breeding effect is remarkable. During the 50 years from 1935 to 1985, the Hawaii Sugar Association of the United States adopted bi-parental hybridization and melting pot hybridization to produce fuzzy. Although sugarcane breeders put more energy into biparental hybridization, the number of hybrid seeds obtained was only 15% of melting pot hybridization. The seedlings obtained by bi-parental hybridization were only 20 % of the total seedlings. As of 1985, among the 10 varieties developed in Hawaii, 9 were developed from melting pot hybridization; 12 sugarcane varieties bred from 1985 to 2005 were all derived from melting pot hybridization. Examples that can be found are 'H62-4671' (Heinz et al., 1979), 'H70-144' (Heinz et al., 1983), 'H65-7052' (Heinz et al., 1981a), 'H68-1158' (Heinz et al., 1981b), 'H73-6110' (Heinz et al., 1984), 'H74-1715' (Tew et al., 1988), 'H74-4527' (Tew et al., 1992b), 'H78-4153' (Wu, 2003) and 'H78-0292' (Tew et al., 1992c) which were bred by melting pot hybridization, and 'H65-7052' is the main variety in Hawaii. Although melting pot hybridization is the primary hybridization method in the commercial breeding of sugarcane in Hawaii, it has not been the primary hybridization method in other countries in the world (Tew et al., 2011).

With multiple hybridization methods, more than half of the cross combinations in the commercial breeding of Philippine Sugar Research Institute Foundation, Inc. are multiple crosses. More than 20 varieties such as 'VMC76-16' and 'VMC88-354' have been developed. Sugar Cane Research Department, Kenana Sugar Company Limited, Sudan developed 'Kn88-260' 'Kn87-65' and other varieties; West Indies Central Sugar Cane Breeding Station in Barbados has bred more than 20 varieties, such as 'B75-466' and 'B01-411'; CIRAD-Guadeloupe has been developed varieties, such as 'Fr89-746' and 'Fr92-394'; and Centro de Tecnologia Canavieira in Brazil bred 'SP80-0185' were all by multiple crossing method (Table 2-7).

Table 2-7 Some sugarcane varieties (lines) bred by multiple crosses in some breeding institutions abroad

Organization	Variety/Clone	Combination	Variety/Clone	Combination
West Indies Central Sugar Cane Breeding Station, Barbados	B01-411	DB85-06 × Polycross	BR03-001	BJ83-124 × Polycross
	B01-424	DB85-06 × Polycross	BR04-002	BJ88-02 × Polycross
	B04-1020	HQ30-41 × Polycross	BR04-003	BJ88-02 × Polycross
	B75-466	B46364 × Polycross	BR04-006	BBZ86-654 × Polycross
	B99-1037	B91-968 × Polycross	BR95-016	UCW54-65 × Polycross
	BJ97-01	BJ88-104 × Polycross	BR98-006	CR68-269 × Polycross
	BJ97-14	BBZ80-19 × Polycross	BR98-24	BJ84-02 × Polycross
	BJ97-19	BJ82-156 × Polycross	DB66113	CB41-76 × Polycross
	BJ97-20	BJ82-156 × Polycross	DB70047	HJ5741 × Polycross
	BJ97-29	R57 0× Polycross	DB96-33	B85-342 × Polycross
	BJ97-36	BJ76-27 × Polycross	DB97-174	DB75-32 × Polycross
	BR00-01	BJ75-55 × Polycross		
(CIRAD) Guadeloupe Experimental Station of CIRAD, France	FG05-424	B831038 × Polycross	Fr95-579	B95579 × Polycross
	Fr89-746	B7352 0× Polycross	Fr98-04	115E × Polycross
	Fr92-394	CP67-412 × Polycross		
Sugar Cane Research Department, Kenana Sugar Company Limited, Sudan	Kn88-260	CO421 × Polycross	Kn94-24	CO421 × Polycross
	Kn87-65	KnH80-540 × Polycross	KnB96-207	B85-788 × Polycross
	Kn88-260	Co421 × Polycross	KnB96-223	B91-1573 × Polycross
	Kn9424	CO421 × Polycross		
Brazil sugarcane industry technology center (Centro de Tecnologia Canavieira)	SP80-0185	BO17 × Polycross		

				continued
Organization	Variety/Clone	Combination	Variety/Clone	Combination
	PSR97-034	VMC82-727 × Polycross	VMC95-09	Q90 × Polycross
	PSR97-051	VMC76-161 × Polycross	VMC95-105	VMC81378 × Polycross
	PSR97-092	VMC51-56 × Polycross	VMC95-110	F108 × Polycross
	VMC73-229	CP38-34 × Polycross	VMC95-119	VMC76-16 × Polycross
	VMC76-16	H28-1813 × Polycross	VMC95-173	F172 × Polycross
Philippine Sugar Research Institute Foundation, Inc.	VMC84-947	POJ2745 × Polycross	VMC95-212	H65-5802 × Polycross
	VMC86-550	P56-226 × Polycross	VMC95-37	M555-60 × Polycross
	VMC88-354	PHIL63-17 × Polycross	VMC95-87	F108 × Polycross
	VMC90-239	57NG67 GreeN × Polycross	VMC95-88	F108 × Polycross
	VMC92-189	VMC68-554 × Polycross	VMC96-134	VMC81-131 × Polycross
	VMC92-228	VMC67-239 × Polycross	VMC96-169	VMC67-273 × Polycross
	VMC93-331	Q84 × Polycross	VMC97-30	Co68-06 × Polycross
	VMC93-339	Q84 × Polycross		

Source: Wu et al., 2014.

Multiple hybridizations are also used in China, but the pollen donors are mostly two male parents. Sugarcane Research Institute, Yunnan Academy of Agricultural Sciences (YSRI) developed 'CYZ 71-998' 'CYZ 72-701' and 'CYZ 83-180'; Guangzhou Sugarcane Industry Research Institute (GZSRI) developed 'CYT 00-319'; Sugarcane Research Institute, Guangxi Academy of Agricultural Sciences (GXSRI) developed 'CGT91-262', Sugarcane Research Institute of Jiangxi Province (JSRI) bred 'CGN 01-129' 'CGN 01-112' and 'CGN 00-378' (Table 2-8). As far as the global sugarcane breeding effect is concerned, the effect of bi-parental hybridization is still the most significant. The main sugarcane varieties 'POJ2878' 'Co419' 'F134' 'CGT11' 'ROC22' 'CGT42' 'CLC05136' 'CYZ0551' 'CYZ081609' and 'CYT 93-159' are developed by bi-parental hybridization, which are widely used.

Table 2-8 Some sugarcane varieties (clones) developed by domestic breeding institutions through multiple crosses

Institutions	Variety (clone)	Cross combination
Sugarcane Research Institue of Guangxi Academy of Agricultural Sciences	CGT91-262	CYT 85-64 × ROC10 + CP49-50
	CGT03-2309	CYT 91-976 × CYT 84-3 + ROC25
Sugarcane Research Institute of Yunnan Academy of Agricultural Sciences and its subordinate breeding stations	CYR 92-148	CGZ 14 × CYN 83-157 + CYN 82-114
	CYR 92-159	CGZ 14 × CYN 83-157 + CYN 82-114
	CYR 92-114	CGZ 14 × CYN 83-157 + CYN82-114
	CYR 92-128	CGZ 14 × CYN 83-157 + CYN82-114
	CYR 92-148	CGZ 14 × CYN 83-157 + CYN82-114
	CYR 93-28	CGZ 14 × CYN 83-157 + CYN82-114
	CYR 04-61	CYR 03-117 × CYR 99-248 + CYR 99-119
	CYR 06-216	H32-5560 × CYR 03-393 + CYR 03-392
	CYR 06-4806	Dezhe 93-94 × CYR 04-186 + CYR 99-155
	CYZ 71-998	Co290 × F180 + CYT 59-264
	CYZ 72-701	CYZ 65-225 × Co331 + CP49-50
	CYZ 83-180	Co419 × CP49-50 + CYC 58-43
Guangzhou Sugar Industry Research Institute and its subordinate breeding station	CYC 57-36	Tanzhouzhucane × F134 + Co331
	CYC 96-64	IJ76-315 × CYC 95-41 + CYC 95-6
	CYC 89-26	F134 × CP72-1210 + CYC 58-47
	CYC 89-35	ROC1 × CP72-1210 + CYC 58-47
	CYC 89-37	CP72-1210 × ROC1 + CYC 58-47
	CYC 89-46	CP72-1210 × CYC 82-108 + CYC 57-25
	CYC 93-2	CP72-1210 × CYC 90-19 + CYC 90-31
	CYC 93-25	CP72-1210 × CYC 90-3 + CYC 88-33
	CYC 93-26	CP72-1210 × CYC 90-3 + CYC 88-33

		continued
Institutions	**Variety (clone)**	**Cross combination**
Guangzhou Sugar Industry Research Institute and its subordinate breeding station	CYC 93-4	CP72-1210 × CYC 90-19 + CYC 90-31
	CYC 93-47	ROC10 × CYC 89-19 + CYC 91-27
	CYC 96-17	CYC 93-25 × CP72-1210 + CYT 85-1805
	CYC 96-18	CYC 93-25 × CP72-1210 + CYT 85-1805
	CYC 96-24	CYT 83-257 × CYC 85-45 + CYC 90-56
	CYT 85-177	CYT 57-423 × CP57-614 + CP72-1312
	CYT 00-319	CYN81-762 × CP72-335 + CP82-1592
Sugarcane Research Institute of Jiangxi Province	CMT 63-41	CHN 53-63 × CYC 58-43
	CGN 00-378	CGZ 14 × CP57-614 + CZZ 82-339
	CGN 01-112	CGN 75-65 × CP57-614 + CZZ 82-339
	CGN 01-129	CGZ 14 × CP57-614 + CZZ 82-339
Sichuan Neijiang Academy of Agricultural Sciences	CTC 8	CHN 53-63 × CP49-50 + CP28-11
	CTC 6	CHN 53-63 × CP49-50 + CP28-11

Source: Wu et al., 2014.

2.3 Sugarcane Selfing and Intra-Crossing and Their Achievement

Since the 1930s, there has been much debate among sugarcane breeders about breeding sugarcane varieties by selfing or intra-crossing. Warner of Hawaii thought that many pairs of quantitative trait genes influenced the same trait in a polyploid, and it wasn't easy to make genes homozygous by selfing. As a vegetatively propagated crop, sugarcane only needs fine individuals and does not need a uniform population, so there is no future for improving sugarcane by selfing. Stevenson of Barbados thought that selfing could increase the purity of genes, and recessive traits could be expressed by selfing, and became stable, heritable traits, so he thought that sugarcane could be improved by selfing. Peng (1990), a famous sugarcane breeder in China, studied that the vigor of the progeny of *S.spontaneum* and close selfing did not deteriorate, and excellent types could be obtained; Selfing progeny of the hybrids with *S.spontaneum* did not deteriorate and could be directly bred into excellent varieties. Peng thought that although the selfing offspring's stalk diameter was thinner, the stalk diameter can be restored to normal if it was hybridized immediately. The selfing offspring had defects, but as parents, the defects can be removed by hybridization. Chu (2000) of Ruili Breeding Station, Yunnan Province, thought promoting recombination of sugarcane genes by selfing and eliminating inferior genes can improve breeding efficiency.

2.3.1 The main lessons of self-breeding

Mai Yandao Shu, a sugarcane breeder in Barbados sugarcane breeding farm, bred 'B3353' 'B3354' and 'B27172' by self-breeding through 'POJ2878', which made his successor Stevenson (1950−1965) exceedingly confident in sugarcane self-breeding and almost lost his mind. It is argued that sugarcane breeding should adopt the hybrid heterosis of corn, first breeding inbred lines, then mating with inbred lines, synthesizing single cross and double-cross as basic hybrid parents to synthesize synthetic hybrid vigor comprehensive characters better. However, Stephenson's luck was not as good as that of his predecessor, Mai Yan Daoshu. Until his retirement in 1965, he did not breed an inbred line or establish an inbred line (Stevenson, 1965). Until 1970, Barbados' cultivated varieties were still 'B49119' and 'B54163' bred by Mai Yan Daoshu's breeding route. Since 1947, Pingtung Sugarcane Breeding Farm in Taiwan has designated special people to do the self-breeding work of 'POJ2878' 'F134' and 'POJ3016'. They have stopped and cannot continue due to the lack of heading or infertility (Luo, 1984; Peng, 1990).

Almost every Sugarcane Breeding Organisation has more or less experience in sugarcane self-breeding. The United States, India, Demerara, Guyana, Argentina, Peru, and Sugarcane Research Institute, Yunnan Academy of Agricultural Sciences of China (Table 2-9) all reported successful breeding of sugarcane varieties. Hainan Sugarcane Breeding Station (HSBS), China, also carried out many self-breeding of cultivated varieties in the 1950s−1970s and created much self-bred germplasm. After establishing the Ruili Breeding Station of Sugarcane Research Institute of

Yunnan Academy of Agricultural Sciences, many self-breeding explorations have been carried out since the 1990s self-breeding materials have been created. Because it is difficult for sugarcane to blossom, most modern sugarcane breeders have not designated a specific variety to breed new varieties by self-breeding. Most of them have better natural flowering conditions. It is random to breed self-bred offspring from the seeds of their female parents.

Table 2-9 Varieties/clone developed from selfing

Variety/Clone	Parental varieties/clone	Location	Type	Derived varietie
27MQ1124	Korpi	Australia	Varieties	Many, such as 58N829 and Q161
B3353, B3354	B (30) L7	Barbados	Varieties	Many, such as B4145 and B45181
Co361	POJ2725	India	Varieties	Many, such as P3247 and Co475
CP1165	Co670	U.S.A	Varieties	Many, such as VMC95-243 and Q165
D1135	D103M	Demerara, Guyana	Varieties	Many, such as H240, H28-4399, and H32-1063, etc
Co670	US1484	India	Varieties	Many, such as F31-963 and 75C35
NA56-79	Co419	Argentina	Varieties	Many, such as RB80-6043 and RB72-5828, etc
US1484	POJ213	U.S.A	Varieties	Many, Co670
P63-32	POJ2725	Peru	Varieties	Many, such as Co1007 and Co997
CYC 54-89	F134	Yacheng, Hainan, China	Clone	Many, CYT 58-1291, CST 71-37, and so on
CYC 55-1	POJ2878	Yacheng, Hainan, China	Clone	Many, CZZ 74-426, CZZ 88-64 and so on
B37172	POJ2878	Barbados	Varieties	No report
Co1307	Co797	India	Varieties	No report
Co508	Co214	India	Varieties	No report

continued

Variety/Clone	Parental varieties/clone	Location	Type	Derived varietie
CP807	POJ213	U.S.A	Varieties	No report
RB80-6043, RB72-5828	NA56-83	Argentina	Varieties	No report
CYZ 71-545, CYZ 71-489	CYZ 65-225	Yunnan, Kaiyuan, China	Varieties	No report
CYC 57-10	CP28-19	Yacheng, Hainan, China	Clone	No report
CYC 76-16	CYT 59-65	Yacheng, Hainan, China	Clone	No report
CZZ 82-339	CZZ 74-141	Zhanjiang, Guangdong, China	Clone	No report
CYR 91-3696, CYR 91-3781	CYR 80-114	Ruili, Yunnan, China	Clone	No report
CYR 93-2418, CYR 96-1458	CYR 80-161	Ruili, Yunnan, China	Clone	No report
CYR 99-634	CYR 93-3148	Ruili, Yunnan, China	Clone	No report
CYR 05-38, CYR 05-39, CYR 05-40, CYR 05-41, CYR 05-42, CYR 05-43	CYR 03-315	Ruili, Yunnan, China	Clone	No report
CYR 91-2868	CYR 80-189	Ruili, Yunnan, China	Clone	No report
CYR 99-711	CYR 95-113	Ruili, Yunnan, China	Clone	No report
CYR 99-695	CYR 96-64	Ruili, Yunnan, China	Clone	No report
CYR 05-31, CYR 05-32, CYR 05-33, CYR 05-34, CYR 05-35, CYR 05-36, CYR 05-37	CYR 99-490	Ruili, Yunnan, China	Clone	No report

Source: Wu et al., 2014.

2.3.2 The main achievements of self-breeding

Not many varieties developed directly through inbred lines at home and abroad, and there are 18 in Table 2-9. However, using inbred lines as parents to cross repeatedly has achieved great success and bred a batch of new varieties (Figure 2-3). 'CYC 54-89' and 'CYC 55-1' developed 'CYT 58-1291' and 'CZZ 72-426' in HSBS, China, and then 'CST 71-37' and 'CZZ 88-64' were developed by further hybridization. In Queensland and Buzacott (1967), the inbred line of *S.officinarum* was used as the male parent, and it was a great success as the source of the high sugar gene. For example, the self-bred progeny of tropical 'Korpi' '27MQ1124', was used as the male parent to cross with 'Co270', and excellent varieties with high sugar content such as 'Trojan' 'Orion' and 'Akbar' were bred (Peng, 1990). 'Trojan' was the main cultivated variety in Queensland in the 1950s–1960s. The Bureau of Sugar Experiment Station of Australia (BSES) used 'Trojan' as the parent and used it repeatedly to cross and develop many superior varieties. 12 of the 30 varieties from 'Q100' – 'Q130' were offspring of 'Trojan', including 'Q100' 'Q102' 'Q105' 'Q106' 'Q109' 'Q111' 'Q115' 'Q118' 'Q120' 'Q122' 'Q124' and 'Q129', accounting for 40% of Q varieties.

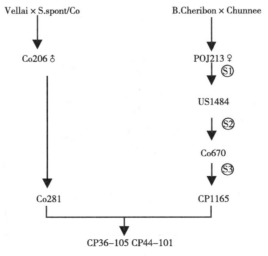

Figure 2-3 Pedigree of CP36-105 and CP44-101

In Fiji, sugarcane's sugar content increased rapidly by crossing between the inbred lines of 'Trojan' and 'Badila'. Barbados has bred 'B3337' and 'B33354' through the inbred line of 'B (30) L7'. The former has production value, and both are good parents. 'B43337' and 'B46364' were bred with the former as parents, and 'B42231' and 'B52107' were bred with the latter. From 1930 to 1960, there were 28 varieties popularized in Louisiana, of which no less than 17 were bred by one inbred line, and 4 were developed by one or two inbred lines (Peng, 1990). Both 'CP807' and 'US1484' are the self-crossed F_1 generations of 'POJ213', and the others are the offspring of 'CP1165' × 'POJ213'. 'CP36-105' and 'CP44-101' (Figure 2-3) are the F_1 generations of the 'CP1165' inbred line and the second and third generation of the inbred line of 'POJ213'. The former has been popularized and applied as the main variety in the state for more than 10 years, while the latter is an excellent variety in Canal Point. Besides, India, Guyana, Argentina, Peru, and other countries have successfully bred self-bred varieties, and excellent varieties have been cultivated through continuous hybridization or backcross utilization of self-bred varieties. It can be seen that variety could be developed through selfing, but there is a low chance for direct selfing. Why all inbred lines, and some varieties have not been bred successfully, but the varieties bred by 'Trojan' are excellent and numerous because 'Trojan' has a similar genetic background to 'Co290' (Figure 2-4), which shows that inbred lines also need to continue to cross and backcross correctly.

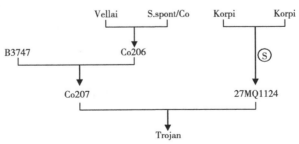

Figure 2-4 Pedigree of Trojan

Ruili National Inland Sugarcane Hybrid Breeding Station (RSBS) has also done much work in self-breeding in recent years. In self-breeding, 8 innovative germplasm were used as a basic parental clone, and more than 20 self-bred offspring were advanced. Because of the short time, no new varieties have been cultivated yet. However, we can learn from previous experiences, continue to cross or backcross in the right way, and expect to develop superior varieties in the future.

2.4 The Method of Selecting Offspring and the Factors Influencing the Selection

The whole breeding process of sugarcane is inseparable from selecting parents, hybrid combinations, hybrid seed cultivation methods, hybrid progenies, test methods, and breeding methods.

In the process of sugarcane breeding, a large number of seedlings are produced every year. When the scale of seedlings is small, the scheme of a high selection rate (more than 10%) is generally adopted to reduce the error of eliminating the excellent individuals who fail to give full play to their production performance at this stage. After years of multi-stage selection, the number of selected clones decreased year by year until completing the multi-point test, regional test and product demonstration, and new varieties breeding. This procedure takes 8–12 years to finish a variety program, which requires many humans, material, and financial resources. Therefore, the selection method is one of the crucial factors affecting breeding efficiency.

Due to the continuous development of sugarcane production and the rise of the sugarcane energy industry, sugarcane production has higher and higher requirements on sugarcane improved varieties. To continuously meet the needs of sugarcane production, in recent years, sugarcane breeders have created more variations and selected excellent varieties by increasing the number of combinations and increasing the scale of seedlings (Figure 2-5). Suppose the approach is still taken to avoid misleading and missed selection. In that case, the scale of sugarcane breeding will continue to expand, human, material, and financial resources will be difficult to bear, so the traditional breeding programs and methods will be difficult to meet the needs of the development of the situation, breeders began to explore new techniques and procedures to improve breeding efficiency.

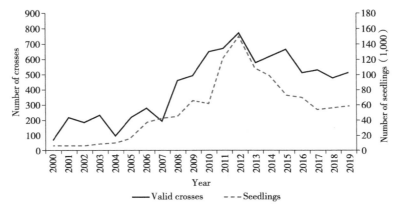

Figure 2-5 Curve of the number of crosses and seedlings in YSRI

Sugarcane varieties usually bear two kinds of selection, namely natural selection and artificial selection. Natural selection is the process of survival of the fittest and elimination of the unfit. Artificial selection refers to selecting the variant types that meet human beings' needs and eliminating those unfavorable to social production under human actions beings. The results of the two kinds of selection are different. The former is to choose the individuals that are beneficial to the survival of the organisms themselves. The latter is to get the individuals who have the characteristics that people expect. These individuals have no advantages in natural competition, but their benefits can be fully exerted in people's environment to meet people's needs.

2.4.1 Selection the method

2.4.1.1 Individual plant selection and family selection

Single plant selection is a method of selecting excellent seedlings according to individual phenotype values. In contrast, the family selection is a method of choosing ideal families according to the average performance of seedlings in the experiment and selecting excellent single plants in deserving families. The selection method mentioned here is mainly for seedling stage selection (Stage 1). Family refers to the seedling population from the same combination.

In Australia, family selection is widely used in the early breeding stage (Stage 1). Since 2005, YSRI has successfully introduced the family selection technology from Australia for the first time. The Yunnan Provincial-Local Standard "Technical Regulations for Evaluation and Selection of Sugarcane Hybrid Breeding Families" has been formed through continuous digestion, absorption, and innovation. The technology has been applied in main sugarcane research institutes in China, and significant progress has been made.

Critical points of family selection: Firstly, the core test is required for the seedlings of all combinations participating in the evaluation. The soil in the core test site is required to be uniform. The seedling of each cross is a randomized block design, and the procedure is repeated three times (if the number of seedlings of some families is not enough, it can be arranged only once or twice repeats). Each plot has 1–2 rows, and each plot is required to plant the same number of seedlings. Simultaneously, the control variety was arranged as the single bud seedling raising could be started about one month before the fuzz seedling transplanting in the field. Secondly, the number of surviving clumps, stalks, height, diameter, and sugar content (or Brix) of each row were collected in the maturing stage. The sugarcane yield and sugar content were obtained by weighed or calculation. Thirdly, the whole family was evaluated according to the collected data (including height, diameter, stalks, cane yield, sugar yield, sugar content, or Brix). Then, individual plants in excellent families were selected according to the evaluation results. Sugar content is the most critical family evaluation index and the most crucial character to calculate families' selection rate. According to the breeding objectives, breeders use other traits to correct the basic selection rate to obtain the actual selection rate and rank the families. In Australia, 50% of families with high selection rates are selected; 70% of the families with higher genetic value in YSRI were selected. In general, the selection rate is generally 20% for the family with the best performance.

The advantages of family selection are as follows: Firstly, the breeding value and genetic value of each trait of the family and parent can be calculated according to the performance of the offspring such as height, diameter, stalks, cane yield, sugar yield, and sugar content (or Brix), and the selection of parents and combinations in the next year can be guided by the calculation results. Secondly, the performance of parents and families and the interaction effect between gene and environment can be evaluated through many years and multi-point experiments. With the increase of using times and years, the evaluation results are more accurate. Thirdly, all the data can be quantified, which is convenient to establish a database and is conducive to computer selection of combinations. Fourthly, the selection rate is based on the genetic value of sugar content of the cross, and the genetic value is the best linear unbiased estimation value. The influence of environmental factors has been eliminated when calculating the genetic value. The selection can avoid the effect of field management quality and human factors, which is more scientific.

2.4.1.2 Direct selection and indirect selection

Whether it is individual selection or family selection, and ultimately need to select in the field directly, the direct choice is the most reliable for the sugarcane breeder. Still, because of its large workload and higher costs, finding the target traits of positively related features and then using their relevance for selection may be the most cost-effective method. Wu et al. (2011) studied the relationship between sucrose content and Brix of four populations of distant hybrid progenies of *S. spontaneum*, with the highest correlation coefficient of 0.985 and the lowest of 0.754, indicating that the correlation between sugar content and Brix was high, which suggested that the selection of sugar content by Brix in the field was effective in the early stage.

Sugarcane quality traits are inherent traits of sugarcane varieties, mainly affected by genetic factors besides environmental factors. The correlation among quality traits can lay a foundation for future sugarcane quality breeding and indirect selection. Research showed (Table 2-10 and Table 2-11) that although there were fluctuations (0.242–0.996) among the quality traits of F_1 and BC_1 from plant and ratoon crops, there were significant positive correlations between them. The variation trend of correlation coefficients of each quality trait was the same in the two populations. It will create conditions for the optimization of sugarcane quality traits and high-quality sugarcane breeding. The results also showed that sugarcane varieties with sugar content and fiber content could be bred by distant hybridization and backcross utilization of wild sugarcane, which created conditions for solving the lodging problem of high yield and high sugar varieties production.

Table 2-10 Correlation analysis of main quality traits both plant and ratoon crop in distant hybrid F_1 of *S. spontaneum*

Item	F_1 plant					F_1 ratoon crop				
	TSH	Fiber	CCS	SCJ	BJ	TSH	Fiber	CCS	SCJ	BJ
Fiber	0.312**					0.255**				
CCS	0.672**	0.246**				0.780**	0.198**			
SCJ	0.684**	0.330**	0.996**			0.787**	0.256**	0.996**		
BJ	0.705**	0.406**	0.910**	0.922**		0.648**	0.333**	0.754**	0.764**	
AP	0.654**	0.274**	0.969**	0.969**	0.814**	0.727**	0.212**	0.941**	0.940**	0.544**

Note: ** The representative significance level was 1%; TSH = tons of sugar per hectare (t/hm^2); CCS = commercial cane sugar (%); SCJ = sugar content of cane juice (%); BJ = Brix of cane juice (%); AP = simple purity (%), he same below.

Source: Wu et al., 2011.

Table 2-11 Correlation between quality traits of plant and ratoon crop of distant crosses BC_1 in sugarcane

Item	BC_1 plant					BC_1 ratoon crop				
	TSH	Fiber	CCS	SCJ	BJ	TSH	Fiber	CCS	SCJ	BJ
Fiber	0.242**					0.457**				
CCS	0.531**	0.296**				0.628**	0.389**			
SCJ	0.531**	0.206**	0.951**			0.652**	0.498**	0.992**		
BJ	0.542**	0.343**	0.985**	0.924**		0.587**	0.485**	0.911**	0.919**	
AP	0.485**	0.379**	0.902**	0.961**	0.788**	0.614**	0.431**	0.940**	0.942**	0.755**

Source: Wu et al., 2011.

How to select, which traits to choose, what kind of selection method to adopt, and how to carry out the early selection are all the problems that sugarcane breeders must explore and solve. In the following circumstances, through predecessors' practice, sugarcane breeding can be carried out by indirect selection.

Firstly, when some traits are highly positively correlated with target traits, and their determination is more convenient, economical, and faster than target traits, breeders can use these traits to select target traits. For example, breeders can choose high sugar varieties by field Brix measurement.

Secondly, some traits are positively related to the target traits. But the natural selection of the target traits in the early stage is too tricky, or the possibility of choice is low, or some data cannot be collected. It is an excellent method to make an indirect selection in the early stage. For example, according to diameter, height, and stalks, high-yield varieties should be selected.

Thirdly, when some early growth stage traits are highly correlated with target traits in the middle and late growth stage, breeders can make the indirect selection in the early growth stage. Such as winter planting sugarcane, according to the emergence of the situation to choose suitable cold resistance varieties; sugarcane harvest after the case of low temperature, according to the number of perennial plant selection of cold resistance, suitable perennial clones; sugarcane harvest drought, according to the number of plants selected strong drought resistance, suitable ratooning varieties.

Indirect selection is generally easy to measure, high efficiency, economically viable, and non-destructive under the premise of use. For example, for the selection of sugar in stage one, due to the limited number of stalks per seedling, it is impossible to take a large number of samples for quality detection. It is challenging to select if a character is too complex to test and the cost is too high, even if it indirectly correlates with the target trait.

Direct selection's genetic progress is more extensive, but the indirect selection can significantly reduce costs. When the heritability of traits used for indirect selection is high, and the genetic correlation between characteristics is high, as shown in Table 2-12, the indirect choice may be more effective than direct selection.

Table 2-12 Examples of indirect selection by correlated character

Correlated character	Target traits
Brix	Sugar content
Diameter and stalks by visual inspection	Cane yield
The hardness of sugarcane skin	The fiber content and lodging resistance
Emergence rate in winter	Cold resistance
Emergence rate in dryland	Drought resistance

Source: WU et al., 2014.

2.4.1.3 Molecular marker-assisted selection

Molecular marker-assisted selection is also an indirect selection technology. With the rapid development of modern biotechnology, molecular marker-assisted breeding in crop breeding is becoming critical. The application scope mainly includes classifying crops, identifying varieties, constructing a genetic linkage map of crops, and the quantitative trait loci location. Molecular marker-assisted selection (MAS) uses molecular markers closely linked to important traits of crops. Molecular markers themselves are not affected by the environment. For characteristics that are easily affected by the environment, molecular marker-assisted selection may improve the efficiency and accuracy of selection, and it is not limited by time and space. The choice is not affected by human factors.

2.4.1.4 The selection of physiological traits

Plant physiology is a science to reveal the essence of plant life phenomenon and to master the knowledge of plant physiology in agriculture. It plays a vital role in production. With the continuous development of physiological research, it is necessary to use physiological characters to study products' quality. More and more breeders pay attention to species selection. Physiological characters are usually helpful to explain the physiological characteristics of crop growth and development.

The physical process and some characters' mechanisms help select sugarcane varieties and develop sugarcane physiological knowledge. The more we know about sugarcane physiology, the more helpful it will be to sugarcane breeding.

The fundamental task of crop breeding is to recognize and transform plants, continuously improve crop yield and quality, robust adaptability, and resistance. Under stress resistance theory guidance, many new sugarcane varieties with strong cold resistance were bred using natural conditions or creating freezing injury conditions. The planting area of Sugarcane was extended from 10° to 30° N or S to 38° N (such as Spain), with an altitude of 1800 m (Yunnan, China). By creating drought resistance, many sugarcane varieties with strong drought resistance have been cultivated under the condition of dry land. The planting area of sugarcane in the dry land of Yunnan Province has reached 80%. The yield per hectare of Sugarcane is still about 60T, and sugar content is still increasing. Many sugarcane varieties with muscular disease and insect resistance were selected by inoculation. The application of pesticides was reduced, and the harm of pests and diseases was continuously reduced. With high fiber wild resources, sugarcane varieties' lodging resistance significantly reduces rodents and disease. The yield and quality of sugarcane were improved. In a word, the objective requirement of sugarcane production and development is essential to plant physiology; a series of urgent research topics have been put forward to inject strong vitality into plant physiology development. Therefore, we can continuously look for problems and absorb nutrients in sugarcane production from plant physiology. At the same time, we should use our research results to solve the major issues in production and continuously strengthen it as the basis of rational agriculture function, which can flourish forward.

Sugarcane is an aneuploid heteropolyploid crop. And its main agronomic and industrial characters are quantitative and controlled by multiple genes. Any trait of sugarcane can not be indirectly selected based on a physiological feature. A physiological character (index) may result from numerous environmental factors, such as drought, increased free proline concentration, and enhanced plasma membrane permeability. Still, high-temperature stress and salt stress can also cause similar changes. It is challenging to investigate the natural state's physiological characteristics in the natural state, such as the varieties with high photosynthetic intensity when using a few plants or leaves as the analysis unit. If there is no right plant type, population structure, and adaptability to specific environmental conditions, the variety may not produce high. The production varieties are usually selected from many clones under the environment of diversity. These varieties generally have good adaptability to the natural conditions and production level similar to the chosen area. The research on variety physiology often lags far behind the breeding of varieties, which indicates that the research level of physiology has not been developed to the stage that can be applied to the effective selection means. According to the production performance of sugarcane varieties, the choice is still the most effective method.

2.4.2 Factors affecting selection

Many factors affect the selection, including genotype-environment interaction (g × E), field management level, competition effect, the number of expected choices, the economic feasibility of options, etc.

2.4.2.1 Interaction between gene and environment

Although different countries and regions have different understandings of the importance of genotype-environment interaction, almost all sugarcane breeding experiments in different regions need to be carried out in various ecological sites. It shows the universality and importance of this interaction. The interaction between gene and environment reduces the selection efficiency and increases the scale and complexity of breeding programs.

There are significant differences in genotype-environment interaction among different sugarcane varieties. Some particular adaptability varieties have great potential in yield and sugar content in some specific areas but have no apparent advantages in other areas. For example, 'CYR 99-601' planted 5 hm^2 of Sugarcane in Baoshan Shangjiang Sugar Mill in 2006, with an average yield of 201 tons per hectare and the highest product of 300 tons per hectare. The sugar content of Sugarcane exceeded 18.70% in March 2007. The sugarcane yield and sugar content have reached a very high level, but other places' performance is not outstanding or even low. Some varieties can be planted in an extensive range of areas, showing good yield stability and strong adaptabilities, such as 'Nco310' 'Co419' 'POJ2878' 'ROC22' 'CGT11' 'CYZ08-1609', and so on, which have been widely planted in many countries or regions.

Because of the interaction between genotype and environment, different regions and breeding organizations need to combine different hybrid combinations and test the soil, representing the local soil and natural conditions, to breed sugarcane varieties suitable for local cultivation. In practice, the interaction between genotypes and locations is more important than that between genotypes and years. The performance of genotypes varies significantly in different places. From the production point of view, we should first emphasize breeding sugarcane varieties in specific areas and environments. Then, we should select sugarcane varieties suitable for different regions of the whole province and the entire country. The factors are not mutually exclusive; In the selection process, if a variety is found to have good performance in a specific area, it should be popularized and applied in the local area. The experiment found that the soil in the same piece of land is often uneven. The fertility difference is extensive; even the performance of the same variety is also very different. Therefore, in the experiment at all levels, even in Stage One or Seedling Stage, random blocks should be used to reduce the environmental error and improve selection accuracy.

In the early stage of sugarcane breeding, the test can only be carried out in one place due to the number of seedlings. At this time, the selection emphasis should be brix, and the selection of yield traits should be relaxed.

2.4.2.2 Competitive effect

In the early stage of sugarcane variety selection, the plot area is usually small. The seedlings with fast growth in the early stage will have a shaded effect on the adjacent plots, affecting neighboring plants' growth. Therefore, when the plot area is small, the yield difference between clones may result from competitive development. At this stage, yield selection should not be too strict to avoid the inappropriate elimination of more stems and slower early growth, which is not suitable for competition with other early growth clones.

With the experiment's upgrading, the plot area expands continuously (such as a four-row plot), and the competition effect among clones is much smaller than that in the early stage of the small plot area. At this time, weighing or measuring yield in the middle two rows can further reduce competition among clones on the product. Still, the error variable will also increase because the plot area of measuring or measuring output is reduced.

2.4.2.3 The purpose of selection and the number of traits expected to be selected

For example, theoretically, when each trait's selection rate is 10%, only ten plants are needed to choose one character to get one ideal individual. If two characters are selected simultaneously, theoretically, only 100 plants are required to get one perfect individual; When three or more traits are chosen simultaneously, hundreds of plants are needed to select the right individual. In the selection, it is generally necessary to comprehensively weigh the economy of multiple traits and sampling methods rather than considering every characteristic separately. For example, breeders often expect to obtain the disease resistance, adaptability, or clustering of wild resources through distant hybridization of Sugarcane. However, less attention was paid to commercial traits. The purpose of commercial crossbreeding is to obtain sugarcane varieties with high yields and high sugar content. The diameter, hair group, and defoliation only modify the selection coefficient, especially in countries where Sugarcane is harvested mechanically.

2.4.2.4 Economic feasibility of choice

Improper selection procedures and methods may significantly increase breeding costs and seriously affect breeding benefits. If we only pay attention to selecting yield and sugar characters and ignore the disease resistance early, the new varieties that have been bred for several years or even more than ten years may be eliminated due to the disease. It is costly to eliminate varieties in the late stage of breeding. In the breeding process, the selection method must be considered to minimize the cost. According to the characteristics of different stages and traits, it is economical and efficient to select related attributes appropriately. The Sugarcane Research Institute of Yunnan Academy of Agricultural Sciences (YSRI) successfully introduced the family selection technology for resistance screening. The requirements for disease resistance were stricter in stage one. For example, when there was a disease in stage one, many seedlings would be eliminated. The rate of selection (%) = the theoretical selection rate% × (100% − 10 × natural incidence rate%). If the incidence rate was over 10%, no matter how the combination was, the selection rate of the whole combination naturally decreased to 0. Brix has a high positive correlation with the sugar content of Sugarcane (Table 2-10 and Table 2-12), and the determination method is economical and straightforward. Brix survey is an effective method for screening high sugar strains in the early stage. The number of stalks and diameter are effective methods for indirectly selecting high-yield sugarcane varieties, often used in selection practice.

References

BERDING N, 1983. Cane flowering studies. BSES Bulletin (1):45-53.

BUZACOTT J H, 1950. Recent trends in the production of cane seedlings in Queensland by the Bureau of Sugar Experiment Stations. South African Sugar Journal (34):721-727.

CHEN R K, LIN Y Q, ZHANG M Q, et al., 2003. Theory and practice of modern sugarcane breeding. Beijing:China Agriculture Press (in Chinese).

CHU L B, 2000. Study on Breeding System of "YN" Sugarcane —— Using "Heterogeneous Compound Separation Theory" to obtain new germplasm with superior high sugar content in Yunnan *S.spontaneum* F_1. Sugarcane, 7 (4):22-33.

HE H, GAN H P, 1990. Storage technology and application of sugarcane pollen. Guangxi Agricultural Science (3):16-17.

HEINZ D J, MEYER H, WU K K, 1979. Registration of 'H62-4671' sugarcane. Crop Science (19):413.

HEINZ D J, TEW T L, MEYER H K, et al., 1981a. Registration of 'H65-7052' sugarcane. Crop Science (23):402.

HEINZ D J, TEW T L, MEYER H K, et al., 1981b. Registration of 'H68-1158' sugarcane. Crop Science (23):402.

HEINZ D J, TEW T L, MEYER H K, et al., 1983. Registration of 'H70-144' sugarcane. Crop Science (23):402.

HEINZ D J, TEW T L, MEYER H K, et al., 1984. Registration of 'H73-6110' sugarcane. Crop Science (24):825.

LI Q W, CHEN X W, HU H X, 1994. A further report on flowering induction of sugarcane. Sugarcane (3):17-21.

LUO J S, 1984. Sugarcane science. Guangzhou:Guangdong sugarcane society.

PENG S G, 1990. Sugarcane breeding. Beijing:Agriculture Press (in Chinese).

STEVENSON G C, 1965. Genetics and breeding of sugarcane. London:Longmans, Green and Co.

TAI P Y P, 1993. Low temperature preservation of F_1 pollen in crosses between noble or commercial sugarcane and *Saccharum spontaneum* L. Sugar Cane (5):8-11.

TEW T L, WU K K, NAGAI C, et al., 1992a. Registration of 'H73-7324' sugarcane. Crop Science (32):281.

TEW T L, WU K K, NAGAI C, et al., 1992b. Registration of 'H74-4527' sugarcane. Crop Science (32):281.

TEW T L, WU K K, NAGAI C, et al., 1992c. Registration of 'H78-0292' sugarcane. Crop Science (32):282.

TEW T L, WU K K, SCHNELL R J, et al., 1988. Registration of 'H74-1715' sugarcane. Crop Science (28):197.

TEW T L, WU K K, SCHNELL R J, et al., 2011. Comparison of biparental and melting pot methods of crossing sugarcane in Hawaii. Sugar Tech, 12 (2):139-144.

WANG J M, 1976. Sugarcane cultivation physiology. Beijing:Agriculture Press (in Chinese).

WANG L P, FAN Y H, MA L, et al., 1999. The research on photoperiod induction flowering of sugarcane and their utilization in hybridization. Sugarcane (3):1-5.

WU C W, PHILLIP J, LIU J Y, et al., 2011. Inheritance of quality traits of the distant crossing between *S.officinarum* and *S. spontaneum*. Journal of Plant genetic resources, 12 (1):59-63.

WU C W, ZHAO P F, XIA H M, et al., 2014. Modern cross breeding and selection techniques in sugarcane. BeiJing:Science Press (in Chinese).

WU K K, 2003. Registration of 'H78-4153' sugarcane. Crop Science (43):431-432.

ZHU D L, 1983. Application of the leaf-cutting method in sugarcane breeding. Sugarcane genetics and breeding. Department of science and technology, Ministry of agriculture, forestry, and fisheries. Sugarcane Research Institute of Fujian Agricultural University.

3 The Techniques of Creating Independent Parent Systems and Breeding Breakthrough Varieties in Sugarcane

Excellent parents are an important prerequisite for sugarcane genetic breeding. Many new sugarcane varieties and new parents have been bred through "POJ" and "Co" (the two important parent systems), which have made great contributions to sugarcane breeding and the development of the sugarcane industry in the world. However, the long-term and high-frequency use of the two parental systems is also an important reason for serious problems such as the narrow genetic basis, networked kinship, and inbreeding of sugarcane parents. The Sugarcane Research Institute of Yunnan Academy of Agricultural Sciences (YSRI) has made full use of new original *Saccharum* species as parents and successfully bred some parents with a new independent consanguinity system using peer-to-peer hybridization years. We have broken through the restriction of kinship between the two parent systems ("POJ" and "Co") and made important progress in the peer-to-peer hybridization of breeding sugarcane new independent parent systems.

3.1 Types and Characteristics of Parental Innovation

3.1.1 Concept of breakthrough sugarcane varieties

Sugarcane variety is the core technology of the cane sugar industry, and sugarcane variety improvement guarantees the development of the sugar industry. Sugarcane variety refers to a sugarcane population with a certain economic value under certain ecological and economic conditions, which has relatively stable heredity and biology and relatively consistent morphology, adapts to certain natural and cultivation conditions. High yield and high sugar content are the basis for fine sugarcane varieties. Breakthrough sugarcane varieties are generally superior to existing varieties in yield, sugar content, ratoonability, disease resistance, insect resistance, adaptability, and other industrial and agricultural traits. They have made breakthroughs in at least 1–2 traits.

Excellent varieties are not invariable. Different ecological environments, historical periods, production conditions, and cultivation systems have different requirements for varieties. Therefore, breakthrough varieties are also not invariable. Excellent sugarcane varieties obtain the highest yield and sugar yield per unit area in a certain region, a certain period, and a certain cultivation system. They are suitable for the requirements of the local ecological environment, cultivation system, and sugar-making process. Breakthrough varieties refer to those that must break through certain regional restrictions and adapt to a wider ecological environment and cultivation system. It is widely accepted by sugarcane farmers and sugar mills and can replace old varieties in a large range and develop excellent varieties in a larger planting area.

3.1.2 Creation of independent parent system

3.1.2.1 Types of parental innovation

Sexual cross breeding of sugarcane is the most commonly used, common, and effective method globally. Among sugarcane varieties bred in China, sugarcane varieties bred through the method account for more than 98% (Wu, 2005; Wu et al., 2014b). Parental innovation is the source of sugarcane variety innovation. Only the breakthrough of parental innovation can bring a breakthrough in sugarcane variety cultivation. The significance of cross breeding is to produce heterosis. The basic reason for heterosis lies in the gene recombination of hybrid offspring. The basic principle of obtaining heterosis is to select excellent hybrid combinations. The heterosis is strong when the genetic relationship is far, the characters are different, the degree of heterogeneity is high, and the genetic basis is rich. Sugarcane centennial breeding has produced many breakthrough varieties and parents, such as 'POJ2878' 'Co419' 'F134' and 'CP49-50', which have made great contributions to the world cane sugar industry and promoted the great development of the cane sugar industry.

Parental innovation is importing new blood relationships or creating new sugarcane parents by hybridization based on the existing parents. According to the differences in blood composition and proportion of parents, parental innovation can be divided into three types: Parental Improvement, New Parental Creation, and cultivation of an Independent Parent System. Parental Improvement improves some bad characters of existing parents by crossing and importing new fewer blood relatives. Its characteristic is that the input characters are individual. Therefore, the improved characters are local, and the genetic basis of varieties changes little. New Parental Creation refers to creating new parents of the original interspecific hybridization outside the existing parental system. Compared with the existing parents, the new parents input at least two or more original *Saccharum* species. That is, there are great differences in genetic basis. A new Independent parent system refers to the parent system independent of the existing POJ and Co systems. The original ancestors are original *Saccharum* species (including *S.officinarum, S.Sinense, S.barberi,* and wild species, the same below) with good commerciality, high yield, high sugar, and unused. The difference in blood relationships between parents and current common parents should be more than 90% (Wu, 2005; Wu et al., 2014b).

3.1.2.2 Methods and characteristics of parent creation

(1) Acquisition of improved parents

Improved parents are the result of parent improvement. Due to some poor characters of existing parents, some characters can be improved by crossing and importing new blood relationships. Its characteristics are based on the existing parents. The method improves the sugar content or yields characters by crossing with high sugar or high-yield varieties and improving adaptability, cold resistance, and ratoonablity by crossing with wild species. Parental improvement is the easiest and simplest method, and it is also the most commonly used method at present. However, because there is no major change in the blood basis, cultivating breakthrough varieties is low.

(2) Creation of new parents

The creation of new parents refers to new original *Saccharum* species hybridization to produce new varieties/parents with better industrial and agricultural characters that can be used for production or continued hybridization. It is easy to obtain a parent that is different from the old parent, but it is difficult to obtain a new parent, especially a parent with a great breakthrough in blood relationship. The characteristics and methods of creating new parents are interspecific hybridizations with original *Saccharum* species completely independent of the existing parent system; The disadvantages are: the number of original *Saccharum* species used in hybridization is small and can not become a parent system, such as 'CYC 58-47'; The advantages are: further hybridization with the offspring of the original *Saccharum* species can cultivate a new parent system, and hybridization with the existing parents can obtain improved parents (for example, 'CZZ 74-141' and 'YC 71-374' are the improved parents formed after hybridization between 'CYC 58-47' and the existing another parent). The limitation is that it is easy to cultivate new varieties by crossing with existing parents (more than 30 varieties have been cultivated with 'CYC 58-47' and its offspring as parents so far). Still, it is difficult to cultivate breakthrough varieties.

(3) Cultivation of independent parent system

The cultivation of an independent parent system refers to cultivating a parent system containing more than 4 new original *Saccharum* species. It is characterized by obtaining high sugar and high yield gene sources from new original cultivated species and discovering new excellent antigens from wild germplasm that have not been explored and utilized; The basis is the noble breeding theory and the practice of breakthrough breeding. After the noble theory was put forward, the parent heterogeneity of all breakthrough parents / varieties (such as 'POJ2878' 'Co419' 'F134' and 'CP49-50') was as high as more than 90 %; The method is as follows: the original ancestor should deal with peer to peer hybridization, continue to hybridize with the two noble F_1 parents, and then obtain more noble F_2 parents or F_3 parents. The blood proportion of every original ancestor is the same; The advantages are: the cultivation process of independent parent system is the process of producing a large number of breakthrough varieties, and the better the original ancestor character, the more likely the offspring will have better varieties; The disadvantages are: it isn't easy and takes a long time. There are a series of problems, such as difficult booting, heading, flowering, poor pollen development, inconsistent florescence, difficult hybridization, low seed setting rate, and poor germination of original excellent *Saccharum* species and their F_1 and F_2 generations, which make it difficult to use a large number of excellent original species.

Misunderstandings in cultivating an independent parent system: Firstly, after hybridization between original *Saccharum* species, the noble F_1 parents are used to cross with another original *Saccharum* species. Although the parents produced by this method integrate the blood relationship of multiple original species, 50% of the blood proportion (Pedigree calculation) is the original species of the last hybridization. For each hybridization of the original species used for the first time, the blood relationship (character) proportion will be reduced by 50%. Secondly, after hybridization between original *Saccharum* species, the noble F_1 or F_2 parents are used to cross with the existing parents. In terms of the phenomenon, the performance of the offspring (F_1 and F_2 generation) produced by the original hybrid may not be as good as the existing varieties/parents, and the improved varieties/parents are easy to be obtained by the hybrid between the original hybrid offspring and the existing parents. However, the blood relationship can only be classified into the old POJ and Co systems after the hybrid. The resulting varieties/parents belong to the improved or innovative type. Because these parents/varieties have a certain cross-blood relationship with existing parents (there is no breakthrough in blood relationship with POJ and Co parent systems), it is difficult for their offspring to make a commercial breakthrough.

3.2 Contribution of Independent Parents System to Sugarcane Cross Breeding

3.2.1 Importance of developing independent parent systems

Parents are an important cornerstone for sugarcane breeding. Only the breakthrough of parents can breed new sugarcane varieties with breakthroughs in variety properties (Chu, 2000; Lang et al., 2015; Luo, 1984; Li, 2010; Chen et al., 2011; Wu, 2005; Wu, 2018; Wu et al., 2014b; Zhuang et al., 2006). Research on how to use the germplasm resources of sugarcane to continuously select and breed good varieties that can meet the production needs, and variety improvement and renewal is the fundamental guarantee for the sustainable development of the sugarcane industry (Li, 2010; Chen et al., 2011; Liu et al., 2003; Wu, 2002; Wu, 2005; Wu et al., 2014a ; Wu, 2014b; Wu, 2018; Zhuang et al., 2006).

Since sugarcane hybridization breeding has been carried out on a large scale, most of the genetic basis of sugarcane breeding in the world can be traced back to the parents of "POJ" and "Co" systems. Eight *S.officinarum*, one *S.Barberi*, and two *S.Spontaneum* were involved. The long-term and frequent use of the two main parental systems of "POJ" and "Co" leads to the narrow genetic basis, networked use, and serious inbreeding of sugarcane parents (Chen et al., 2011; Wu, 2005; Wu, 2018; Wu et al., 2014a; Wu et al., 2014b; Zhuang et al., 2006).

Yunnan has built the unique National Germplasm Repository of Sugarcane in China. The collection from 34 countries and China's 12 provinces, 6 genera, and 15 species cataloged sugarcane germplasm resources. 3,151 germplasm resources after the world's largest Sugarcane Germplasm Resources Conservation Center of India's 5,143 (Mirjkar et al., 2019) is China's largest and the second-largest sugarcane gene bank the world.

There are so many sugarcane germplasm resources, and the level of research technology has been continuously improved. However, sugarcane parents' narrow genetic basis, Networked kinship, and inbreeding of sugarcane parents have not been solved for a long time. We studied the hybrid utilization methods and effects of sugarcane germplasm resources and found that the contributions to sugarcane breeding were different due to the different hybrid methods (Chen et al., 2011; Wu, 2005 ; Wu, 2018; Wu et al., 2014b).

3.2.2 Contribution of the independent parent system to sugarcane breeding by using the basic germplasm for peer-to-peer hybridization

Peer-to-peer hybridization is defined as a hybridization in which the number of original *Saccharum* species and the number of genetic generations contained in the male parent and the female parent of a hybrid combination are the same. The original *Saccharum* species are in the blood source of the parent is not repeated or intermingled. Peer-to-peer hybridization is the hybridization between the original *Saccharum* species and the original *Saccharum* species to produce the noble F_1 generation variety, which can be further carried out in more generations of peer-to-peer hybridization to obtain the noble F_2, F_3, or higher generation variety. The 100-year breeding experience of sugarcane has proved that the breakthrough in parental consanguinity lies in parental heterogeneity and cross symmetry (Wu, 2005; Wu, 2018; Wu et al., 2014b). For example, the National Sugarcane Breeding Farm of Java used S. officinarum for peer-to-peer hybridization. They used Lahaina and Fiji to hybridize and breed F_1 generation variety 'EK2'. B.itan and Loethers were also used to hybridize and breed F_1 generation variety 'POJ100'. Then 'EK2' was used as the female parent and 'POJ100' as the male for peer-to-peer hybridization to obtain the F_2 generation 'EK28'. Since there was no kinship crossing between 'EK2' and 'POJ100' (Figure 3-1), 'EK28' had a greater yield and sugar content than the 4 original parents Lahaina, Fiji, B. hitan, and Loethers, even the F_1 generations 'EK2' and 'POJ100'. The development, production, and application of 'EK28' greatly contributed to the Javan sugar industry in the transitional period (Luo, 1984; Wu, 2005; Wu et al., 2014b).

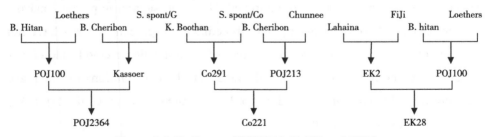

Figure 3-1 Pedigree of POJ2364, Co221 and EK28

For another example, an F_1 hybrid 'Kassoer' was bred by 'B. Cheribon' (*S. officinarum* L) and *S. Spont* /G (*S. Spontaneum*) at the National Sugarcane Breeding Farm Java. Then, 'POJ2364' of the F_2 generation was bred by peer-to-peer hybridization using 'POJ100' as a female parent and 'Kassoer' as a male parent. Peer-to-peer hybridization between 'B. Cheribon' and Chunnee (*S. barberi*) was carried out, and the F_1 hybrid 'POJ213' was also bred.

An F_1 hybrid 'Co291' was developed by crossing K.Boothan (*S. officinarum*) with S.pont/Co (*S.spontaneum* in India) in the Coimbatore Sugarcane Breeding Farm in India. Then 'Co221' was bred by hybridization using 'Co291' as female parent and 'POJ213' as the male parent. Because there is no consanguinity crossing between POJ2364's two parents 'POJ100' and 'Kassoer', and Co221's two parents 'Co291' and POJ213's two parents (Figure 3-1), hybridization heterogeneity is large, hybridization symmetry, heterosis of hybrid progeny is strong. The blood input of wild species (*S.pont* /G and *S.pont*/Co) is added. 'POJ2364' and 'Co221' showed better performance than their parents and significantly improved stress resistance and adaptability (Luo, 1984; Wu, 2005; Wu et al., 2014a). The parental heterogeneity is great. The pedigree looks like an inverted equilateral triangle, the hybridization is symmetrical, the kinship is clear, and the heterosis of the offspring is strong. As the parent, many varieties, excellent characteristics, and many worldwide varieties are produced by the offspring, making great contributions to global breeding (Wu, 2005 ; Wu, 2018; Wu et al., 2014b).

3.2.3 Contributions of "POJ" and "CO" parent systems to sugarcane breeding and cane sugar industry in the world

Breeding of breakthrough sugarcane varieties depends on parental consanguinity, and parental innovation is the key (Chen et al., 2011; Wu, 2005; Wu et al., 2013; Wu, 2018). Based on the two-parent systems represented by 'POJ2878' and 'CO290', sugarcane varieties and parents such as 'CO419' 'F134' 'CP49-50' and 'NCO310' have been bred for a hundred years through crossbreeding, which has made significant contributions to the development of the world sugar industry and promoted the great development of the sugar cane industry (Chu, 2000; Lang et al., 2015; Luo, 1984; Li, 2010; Chen et al., 2011; Wu, 2005 ; Wu, 2018; Wu et al., 2014a; Wu, 2014b; Zhuang et al., 2006).

For example, the parent of 'POJ2878' was 'POJ2364' × 'EK28', and there were 6 original *Saccharum* species. Both parents contained B. hitan and Loether, and the other 4 were completely different (Figure 3-2), with the heterogeneity of 66.7%.

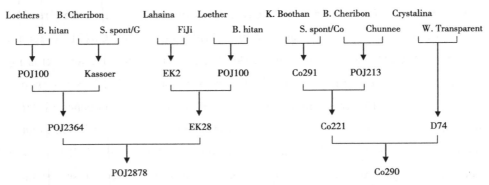

Figure 3-2 The pedigree of POJ2878 and Co290

Another example is that the parent of 'Co290' is 'Co221' × 'D74', which also involves 6 completely different *Saccharum* species, and there are no identical *Saccharum* species in the pedigree (Figure 3-2), showing 100% heterogeneity. According to the analysis of the blood components of 'POJ2878' and 'Co290', there were 11 original *Saccharum* species of 'POJ2878' and 'Co290'. The *S.officinarum* 'B. Cheribon' was the common parent 'POJ2878' and 'Co290', and the heterozygosity was 90.9%. They crossed positively and inversely and bred 'Co419' and 'F134' and many other worldwide varieties and parents (Chen et al., 2011; Wu, 2018; Wu et al., 2014b).

It has laid the foundation of two major sugarcane breeding systems, "POJ" and "Co" globally. The only defect of the two-parent systems is that 33.3% of the original *Saccharum* species of 'POJ2878' have crossed bloodlines. In comparison, the heterogeneity of 'Co290' is 100%, but the cross blood symmetry was slightly poor (Wu et al., 2014b; Wu, 2021). As the reciprocal progeny of 'POJ2878' and 'Co290', 'Co419' and 'F134' have more excellent species and adaptability than 'POJ2878' and 'Co290' because they contain a large number of original species, wide sources of excellent genes, and high heterogeneity. These varieties and parents showed strong heterogeneity and cross symmetry (Wu et al., 2014b). Their blood relationship is clear; their offspring have strong heterosis, as production varieties, they have excellent performance, great breakthrough, large promotion area, and great contribution to the cane sugar industry (Li, 2010; Chen et al., 2011; Wu et al., 2014a; Wu, 2014b; Wu, 2021; Zhuang et al., 2006).

3.3 Development of Creating New Types of Independent parent Systems by Peer-to-Peer Hybridization in Yunnan

3.3.1 The idea of peer-to-peer hybridization to breed independent parent system

The parents are the material foundation stone for breeding breakthrough sugarcane varieties (Chen et al., 2011; Wu, 2005 ; Wu, 2021; Wu et al., 2014b). Over the past ten years, sugarcane crossbreeding workers worldwide have done a lot of work in germplasm innovation and crossbreeding utilization of parents. Thousands of hybrids have been selected, and a large number of parents have also been created. However, because the hybridization has not been separated from the two-parent systems of "POJ" and "Co" , As a result, there was no significant breakthrough in the traits of sugarcane varieties developed later.

Why are there many sugarcane germplasm resources, and the problems of sugarcane parents' narrow genetic basis, networked kinship and inbreeding of sugarcane parents, have not been solved for a long time? Since the 1990s, domestic sugarcane breeding institutions have carried out extensive scientific and technological cooperation with major global sugarcane scientific research organizations. Some scientific and technological personnel have successively gone to major sugarcane-producing countries and organizations in the world for investigation, study, and cooperative research and have mastered much first-hand information on sugarcane crossbreeding (Wu, 2005; Wu, 2021).

Based on absorbing the experience and lessons of sugarcane hybrid breeding at home and abroad, adjusting the breeding ideas, and innovating the breeding methods, the Sugarcane Research Institute of Yunnan Academy of Agricultural Sciences (YSRI) for the first time proposed the idea of using the independent parent system to cultivate breakthrough sugarcane varieties through peer-to-peer hybridization (Wu, 2005; Wu, 2021; Wu et al., 2013).

3.3.2 Utilization of new *Saccharum* species in basic hybridization

Yunnan is the second-largest cane sugar production base in China and one of the regions with the richest wild sugarcane resources globally (Fan et al., 2001; Bian et al., 2015; Liu et al., 2014; Tao et al., 2011). To promote the collection and innovative utilization of sugarcane germplasm resources, China has successfully established the National Germplasm Repository of Sugarcane and the Ruili Inland Hybrid Breeding Base in Yunnan 1980s. By the end of the 13th Five-Year Plan of China, relying on the National Germplasm Repository of Sugarcane, YSRI has established the Sugarcane Germplasm Resources Database System, which has the largest number of sugarcane resources, and the complete data in China, and has cataloged 3,151 sugarcane resource materials from 15 species of 6 genera. Phenotypic traits combined with molecular biological technology, a batch of sugarcane genetic materials with excellent characters such as sugar content, yield, disease resistance, and drought resistance were explored and excavated. On the use of germplasm innovation according to the thinking of cultivating new independent parent system, through the resistance evaluation on a large scale, flowering characteristics, and photoperiodic induction research, successively using 21 new original *Saccharum* species (including *S.officinarum*, *S. sinense*, *S.*

barberi, the endemic species, *S.spontaneum*, *S. robustum*) (Table 3-1), to launch a new basic hybridization between original *Saccharum* species. Some excellent basic hybrid germplasm of F_1 generation were created, laying a solid foundation for the cultivation of new independent parent system (Jing et al., 2013; Tao et al., 2014; Bian et al., 2014; Tian et al., 2017).

Table 3-1 Summary of new independent parents bred by Yunnan peer-to-peer hybridization using new *Saccharum* species

Sugarcane variety type	Name of sugarcane variety	Quantity
S. officinarum L.	48Mouna, 51NG90, Badila, Barwilspt, Zopilata, Canablanca, Ganarafy, 50uahapele, Luohan cane	9
S. sinense Roxb.	Pansahi, Bailou cane, Guangze bamboo cane, Xuchang Henan chewing cane, Nanjian chewing cane	5
endemic species	Baimei cane, Pupiao chewing cane	2
S. robustum Brand et. Jewiet	51NG63, 57NG208	2
S. spontaneum L.	Yunnan82-59, Yunnan 82-114, Yunnan 82-157, Yunnan 84-268	3

Source: Wu, 2021.

3.3.3 Development and advantage analysis of new independent parent system by peer-to-peer hybridization

By interspecific hybridization of the original *Saccharum* species, YSRI carried out the number of peer-to-peer hybridization generations. A great breakthrough has been made in cultivating a new independent parent system by breaking the narrow consanguinity basis of existing sugarcane varieties, Networked kinship, and the inbreeding of sugarcane parents. A batch F_1 generation excellent innovative germplasm has been created. Through peer-to-peer hybridization, 'CYR13-47' 'CYR 14-211' 'CYR 15-55' 'CYR 17-82' and 'CYR 18-188' were obtained a batch of new F_2 and F_3 parents with a notable breakthrough in traits (Tao et al., 2020; Yu et al., 2019a; Yu et al., 2019b; Tao et al., 2015; Hu et al., 2021), and the pedigree of some parents with independent blood system was shown in Figure 3-3.

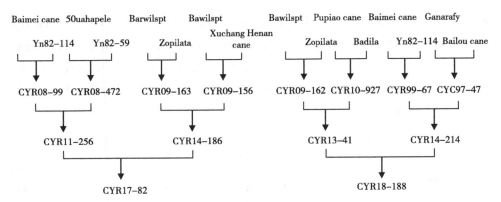

Figure 3-3 The pedigree of CYR17-82 and CYR18-188

Note: Yn82-114 and Yn82-59 are the numbers of *S.Spontaneum* in Yunnan.

Compared with 'POJ2878' and 'Co290' parent systems, the advantage of the new parent system was that all the hybrid parents were new original *Saccharum* species, and the number of new original *Saccharum* species used was 21, which was more than the sum of the two-parent systems of "POJ" and "Co" . At the same time, the problem of 'POJ2878' crossover was overcome, and the poor symmetry of 'Co290' hybridization was avoided.

The results of field experiment observation and hybrid progeny showed that these parents had high yield, high sugar content, strong Ratoonability, excellent disease resistance, and wide adaptability. The main characteristic was good heritability. They can be used as a parent to select hybrid combinations with existing 'POJ2878' and 'Co290' parent systems and offspring-derived varieties. Due to the lack of cross relatives and the strong heterosis of cross hybrids is expected to become the main source for breeding breakthrough sugarcane varieties in China and even globally. The first batch of new parents with F_2 generation of independent parent system has been provided to the Hainan Sugarcane Breeding Station (HSBS), China, in 2019. The catalog is shown in Table 3-2. After the study of flowering characteristics is completed, the next step is expected to gradually provide the fuzz of the new hybrid progeny to sugarcane breeding organizations nationwide.

Table 3-2 The list of parents to HSBS in 2019

NO.	Parents and sources	Type	Quantity
1	CYR15-55 (CYR09-170 × CYZ03-232), CYR14-168 (CYC58-47 × CYR09-173), CYR14-178 (CYR09-160 × CYR09-16), CYR14-211 (CYR09-67 × CYC97-47), CYR13-47 (CYR 09-169 × CYR10-915)	F_2 generation by peer-to-peer hybrid	5
2	CYR 15-90 (CP94-1100 × CYR11-103), CYZ16-1001 and CYZ16-1002 (CYT 93-159 × CYR10-688), CYZ16-1005, CYZ16-1006, and CYZ16-1007 (CYR13-46 × SP80-0185), CYZ16-1008 (CYR14-192 × CYZ05-49)	F_2 generation with the consanguine of peer-to-peer hybrids	7

Source: Wu, 2021.

3.3.4 Problems in the system of peer-to-peer hybridization of sugarcane independent parents

The process of sugarcane crossbreeding utilizes heterosis, which refers to the phenomenon that F_1 generation hybrids produced by the cross of two parents with different traits are significantly superior to their parents in yield, quality, disease resistance, adaptability, and other aspects. Parents with more excellent traits, less cross kinship, strong heterogeneity, and good combining ability will have more excellent offspring. It is easier to breed breakthrough varieties (Wu, 2005 ; Wu, 2021; Wu et al., 2014b).

The practice has proved that parents with independent consanguinity have good genetic characteristics and outstanding super-parent advantages and breed breakthrough sugarcane varieties. The main problems existing in the parent system of breeding new sugarcane independent consanguinity through reciprocal hybridization in Yunnan are as follows: Firstly, the most important donors with large stems, high juice, low fiber, sugar content, and resistance to diseases such as smut are *S.officinarum*, whose origin centers are located in eastern Indonesia and New Guinea of the Pacific Ocean (Luo, 1984; Li, 2010; Chen et al., 2011; Peng 1990; Wu et al., 2014b). *S.officinarum* resources are very scarce in the National Germplasm

Repository of Sugarcane (NGRS), China. Secondly, the number of other original cultivated species (including *S. sinense* Roxb., *S.barberi*, and endemic species) collected in NGRS is also less. The number of originals with excellent performance is not much. Thirdly, it is difficult for the original species to blossom and flowering synchronization, and the success rate of hybridization is low. In creating new independent parent systems, it is not easy to make sure that all the original species can be effectively used as we designed, that the original species of the same parent do not repeat, and that the traits are completely complementary in selecting and matching. For example, Bawilspt, one of *S.officinarum* in 'CYR 17-82' parents, was used twice. At the same time, it also contains the consanguinity of 'Yn82-114' and 'Yn82-59' (*S. spontaneum* L.), resulting in the consanguinity of wild species as high as 25%. Fourthly, some parents contain the relatives of *E. arundinaceum* (Tao et al., 2014; Tao et al., 1997). *E. arundinaceum*, as a relative sugarcane species, has not been bred into varieties after decades of cross-utilization at home and abroad, and the effect of improving sugarcane varieties needs to be further verified. Fifthly, in theory, 8 original species are needed to breed a good new independent consanguineous F_3 parent. At least 7 original cultivated species (including *S.officinarum*, *S.sinense*, *S.barberi* J., and endemic species) can be used as the female parent. There are only 16 cultivated original species in use at present, and at most, two parental systems with completely different blood ties can be created. Therefore, blood ties crossing among different parents is inevitable. For example, 'CYR 17-82' and 'CYR 18-188' have no duplication with the two independent parent systems of "POJ" and "Co", but they have the same four original species of Baimei cane, 'Yn 82-114', 'Barwilspt', and 'Zopilata' (Figure 3-3).

3.4 Genetic Relationship and Breeding Effect of Sugarcane Backbone Parents in China

Parent resources are the foundation of sugarcane breeding. The germplasm is the carrier of excellent gene resources. Without good germplasm resources, it is hard to cultivate excellent parents. Accordingly, it is challenging to breed breakthrough varieties without the breakthrough of parent materials. According to the incomplete statistics, 124 parents used it effectively, and 300 new varieties were bred since the beginning of sexual hybridization breeding in China. It was noted that 10 backbone parents had been used for breeding 144 varieties by direct and continued hybridization. Thus it can be concluded that creating and screening backbone parents for hybridization will significantly improve breeding efficiency.

3.4.1 Genetic relationship and breeding effect of ten backbone parents

Since the sexual hybridization of sugarcane, the excellent F_1 generation obtained through the basic cross of the original *Saccharum* species was continued to peer-based hybridization, resulting in excellent parent materials. These materials exhibited clear genetic relationships, easy flowering, good hybrid affinity, and strong heterosis, considered backbone parents. Table 3-3 showed the top ten backbone parents with the most varieties which were bred since half a century of sugarcane sexual hybridization in mainland China. It was found that the basic hybrid modes were all interspecific hybridization. There were more types of symmetric hybridization than asymmetric hybridization in the breeding program. There was less genetic crossover in consanguinity proportion. The breeding efficiency of backbone parents was very high. It was found that the top ten backbone parents accounted for only 8.1% of the 124 used parents, but the bred varieties accounted for 58.5% of the total. However, the aggravating trend of aging parents brings an impact to sustainable sugarcane development. Eight of 10 backbone parents are 'F134' 'CP49-50' 'Co419' 'F108' 'CP28 53-63-11' 'CHN53-63' 'Neijiang57-416' ('CNJ57-416') 'NCo310' used to crossbreed for almost half a century. No new variety was bred in the last ten years, and no excellent offspring was developed in the last 30 years. The innovation of parents was not enough in China. Seven of 10 backbone parents were introduced from abroad, which bred 120 new varieties with an average of 17 varieties per parent. There were only 3 innovative parents cultivated in China, and 20 varieties were bred with an average of 7 varieties per parent. Therefore, the number of backbone parents cultivated in China is less and their breeding efficiency is relatively lower.

Table 3-3 Statistics on the utilization effect of top ten backbone parents in China

No.	Parents	Year	Female	Male	Number of bred varieties
1	F134	1954–1977	4	39	43
2	CP49-50	1957–1986	12	30	42
3	Co419	1959-1999	26	4	30
4	CP72-1210	1989–1903	14	7	21
5	NJ57-416	1966–1982	3	12	15
6	NCo310	1961–1979	15	0	15
7	F108	1954-1971	10	3	13
8	CP28-11	1958–1971	0	12	12
9	CHN53-63	1958–1967	10	1	11
10	CYC71-374	1989–1996	2	9	11

Source: Wu et al. 2014b.

'F134': The hybrid combination is 'Co290' × 'POJ2878', both of which are worldwide parents. It was bred by Taiwan Sugarcane Industry Research Institute, China (TWSRI), in 1945. This variety has a clear genetic relationship. It had been widely promoted in China since it was introduced in 1947 and had been widely used as a hybrid parent from 1954 to 1977. In the past 30 years, it bred into the most sugarcane varieties in China, with as many as 43 direct offspring, accounting for 38.9% of all bred varieties at that time. More importantly, the offspring varieties 'CNJ57-416' 'CHN 56-12' and 'CYT59-65' are excellent parents. The varieties with the consanguinity of 'F134' account for almost 50% of the total 240 varieties bred in China up to 2010. Due to repeated hybridization and utilization, it is challenging to create and explore symmetrical parents, and the genetic relationships of offspring varieties are closely related. In the late 1980s, the number of breeding varieties directly used as hybrid parents was significantly reduced (Figure 3-4).

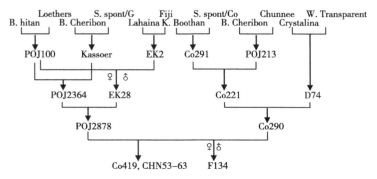

Figure 3-4 The pedigree of Co419, CHN53-63 and F134

'CP49-50': The hybrid combination is 'CP34-120' × 'Co356'. 'Cp49-50' is one of China's most commonly used parents from the late 1950s to the 1990s (1957–1986). Through direct hybridization utilization, 42 sugarcane varieties were bred. Its offspring varieties, 'CGT11', used to be the main varieties with the most prominent promotion area and the most significant economic benefits in China. After the 1990s, the direct hybridization of this parent had not produced new varieties. Instead, many of its offspring became important parents, such as 'CYN73-204' 'CGT73-167' etc. Their pedigree charts are shown in Figure 3-5.

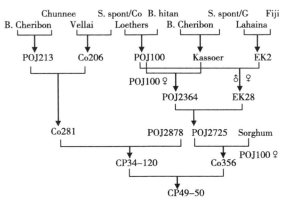

Figure 3-5 The pedigree of CP49-50

'Co419': The hybrid combination is 'POJ2878' × 'Co290'. It shows the advantages of yield and adaptability and was introduced early in India. 'Co419' and

'F134' have the same parents, but the manner of parents cross opposite. Like 'F134', 'Co419' is also a very nice parent. Thirty varieties and parents have been directly bred by hybridization utilization of 'Co419' in China, including 'CGT11', 'CYZ 65-225' (Figure 3-4).

'CP72-1210': The hybrid combination is 'CP65-357' × 'CP56-63'. 'CP72-1210' was bred in the Canal Point Sugarcane Breeding Station of the United States (CPSBS). This variety once accounted for more than 60% of the acreage in Florida due to its excellent traits of millable stalks and strong ratooning (Chen et al., 2011). Since the 1980s, 'CP72-1210' has become the most commonly used parent of sugarcane crossbreeding in mainland China due to its strong combining ability and high sugar content. There were 20 fine varieties origin from 'CP72-1210', which included 'CFN91-4621' 'CYT00-236' and 'CYT93-159'. It was found that 'CP72-1210' has serious networking and poor symmetry in the consanguinity relationship (Figure 3-6). However, 'CP72-1210' showed high breeding efficiency as a parent, particularly among the top ten backbone parents. But 'CP72-1210' was bred so early. It is not known whether each generation is a true hybrid. Another reason may be that it contains the consanguinity of 'CP1165', which eliminated unfavorable genes by three times of self-pollination.

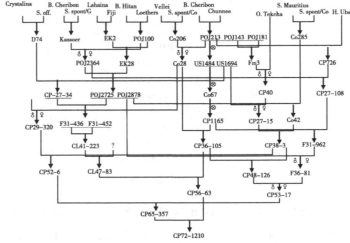

Figure 3-6 The pedigree of CP72-1210

'CNJ57-416': Its parental system is 'Co281' × 'F134'. It is one of the excellent sugarcane parents from the 1960s to 1980s in China. Its maternal parent, 'Co281' belongs to the F_3 hybrid of *S.officinarum*, *S. barberi* (Chunnee), and *S.spontaneum*. The offspring 'CNJ57-416' exhibited solid cold resistance and a fast growth rate. During 20 years, fifteen sugarcane varieties were cultivated in the north sugarcane area of China (Jiangxi, Sichuan, Hunan, etc.), among which 'CCZ17' and 'CYZ89-151' were widely planted in Sichuan and Yunnan provinces, respectively. As Figure 3-7, it can be seen that the genetic relationship of 'CNJ57-416' is clear. But the symmetry of hybridization is insufficient, which is the limitation of the parent utilization.

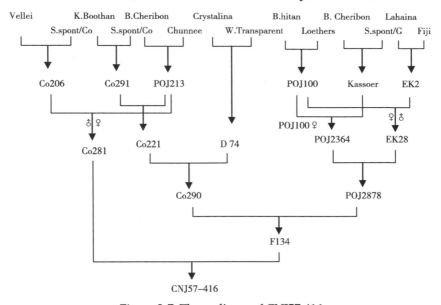

Figure 3-7 The pedigree of CNJ57-416

'NCo310': The parental cross is 'Co421' × 'Co312'. It is one of the excellent female parents in the 1960s and 1970s in China. During 20 years, twenty sugarcane varieties have been cultivated in Jiangxi, Sichuan, Guizhou, and Hunan. Among their offspring, 'CCZ 13' has been widely promoted in Sichuan and Yunnan sugarcane areas (Figure 3-8).

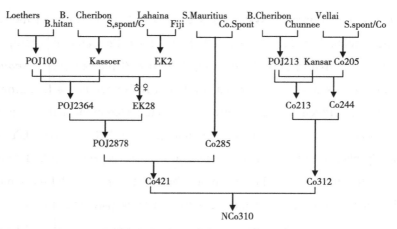

Figure 3-8 The pedigree of NCo310

'F108': The parental cross is 'POJ2725' × 'F46'. 'F108' bred in TWSRI, exhibited precocious, high sugar content and was strongly resistant to wind. It was introduced to mainland China in 1947 and mainly planted in good water and fertilizer conditions in Guangdong, Guangxi, Fujian, and other provinces. It was the primary hybrid parent from the 1950s to 1960s in China. Many excellent varieties had been bred. Its derived varieties were also important parents, including 'CHN56-12' 'CHN56-21' 'CYT57-423' and 'CYT85-177' (Figure 3-9).

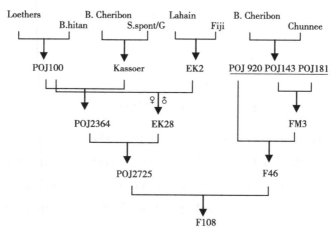

Figure 3-9 The pedigree of F108

'CP28-11': The parental cross is 'Co281' × 'US1694' (Figure 3-10). It was bred in CPSBS and was introduced to China in 1937. It was mainly distributed in the northern margin of Sichuan. This variety has many advantageous traits with early maturity, high sugar content, numerous millable stalks, well-grown but exhibits small stem diameter, scattered architecture, and accessibility to lodging. Its offspring grew fast and exhibited good cold resistance. It was an excellent male parent from the late 1950s to the early 1970s in Sichuan Province. Twelve sugarcane varieties were bred in Sichuan Sugar Industry Research Institute and Sichuan Neijiang Agricultural Science Research Institute.

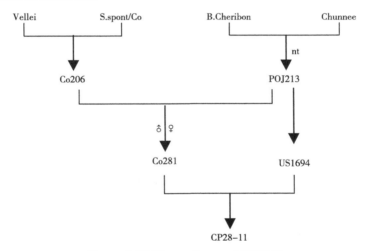

Figure 3-10 The pedigree of CP28-11

'CHN53-63': The parental cross is 'POJ2878' × 'Co290' (Figure 3-4). Guangdong Academy of Agricultural Sciences bred 'CHN53-63'. Its offspring grow faster and have good cold resistance. In the late 1950s and early 1960s, eleven sugarcane varieties were cultivated in Sichuan and Jiangxi using 'CHN53-63' as a hybrid parent.

'CYC71-374': Its parental cross is 'CYT54-143' × 'CYC58-47', and it was bred in HSBS. It is an innovative parent and belongs to the F_2 hybrid of Yacheng *S.spontaneum* and 'Badila' (*S. officinarum*). As the hybrid parent, it was directly used to breed 11 varieties in the major sugarcane regions of China. As shown in

Figure 3-11, 'CYC71-374' is an innovative parent with a clear genetic relationship. But there is a severe asymmetry of hybridization mode. So the breakthrough of the traits is limited in the offspring of 'CYC71-374'.

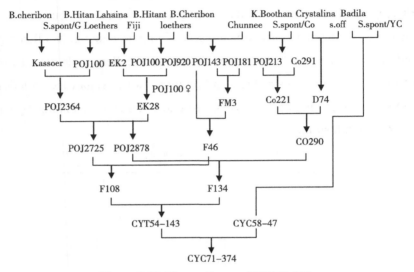

Figure 3-11 The pedigree of CYC71-374

3.4.2 Genetic relationship and breeding effect of ten new sugarcane parents

Because sugarcane parents are used repeatedly, good gene resources have been explored continuously. The creation and discovery of new backbone parents will significantly improve the benefit of sugarcane breeding. Chinese sugarcane breeders have done a lot of work and made remarkable achievements in exploring new parents in recent years. Table 3-4 shows 10 parents that have been widely used with good breeding effects in recent years. By analyzing the new sugarcane breeding parents, the following characteristics can be obtained. Firstly, The ROC series varieties have been gradually changed from production utilization to hybrid utilization. These ROC varieties occupied half of the new top ten parents. Secondly, the breeding efficiency of innovative parents is gradually improved. Since the

1990s, it was found that 49 sugarcane varieties had been bred from 10 new parents, among which 34 varieties (including reciprocal hybrids, the same below) were bred from 7 introduced parents and 18 from 3 innovative parents. The breeding effect of each innovative parent is higher than that of introduced parents. Thirdly, the new top ten parents use 14 original *Saccharum* species, precisely the same as ten old backbone parents. But there is a certain degree of networking, and the symmetry of hybridization is generally worse than that of the ten old backbone parents. Therefore, there is no breakthrough in the breeding effect and the production performance in the offspring of the new parents. So they are called the top parents rather than the backbone parents. Fourthly, to the continuous and stable development of the cane sugar industry, it is urgently needed to use the new germplasm of *Saccharum* species to develop a new independent parent system.

Table 3-4 Improvement and hybrid utilization of new 10 sugarcane parents in China

No.	Parents	Year	Female	Male	Number of bred varieties
1	CP84-1198	1999–2013	1	8	9
2	ROC1	1991–2013	4	4	8
3	CYN73-204	1989–2013	7	1	8
4	ROC10	1995–2013	4	3	7
5	ROC22	2002–2013	1	6	7
6	ROC25	1999–2013	4	2	6
7	Ke5	1991–2013	0	5	5
8	CYT85-177	1996–2013	4	1	5
9	CYT91-976	2002–2013	5	0	5
10	ROC23	2002–2013	2	1	3

Source: Wu et al. 2014b.

'CP84-1198': The parental cross is 'CP70-1133' × 'CP72-2086'. It was bred in CPSBS. It is a medium-large stem, easy to booting, high heading rate, high pollen development rate, a large amount of pollen, suitable for both male and female parents. In contrast, the breeding effect of the male is better. There are 9 excellent offspring bred in mainland China since 1999. 'CP84-1198' shows serious networked kinship (Figure 3-12). Like 'CP72-1210' 'CP1165' was also one of its parents, bred by three rounds of self-pollination. The probability of breeding excellent variety was improved as some unfavorable genes were eliminated in 'CP1165'.

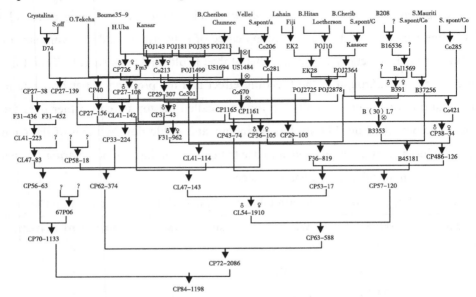

Figure 3-12 The pedigree of CP84-1198

'ROC1': The parental cross is 'F146' × 'CP58-48'. It was bred in Taiwan Sugarcane Industry Research Institute, China (TWSRI). It shows early maturing, high yield, high sugar content, fast growth, drought resistance, water-logging resistance, saline-alkali resistance, and wind resistance. But it is sensitive to low temperatures. It is the important parent for breeding the early maturing and high-sugar-content varieties in China. 'ROC1' is widely planted in Guangdong, Guangxi, Hainan, Fujian, and Yunnan provinces. It began to use 'ROC1' as both male parent and female parent in the early 1990s. Eight sugarcane varieties have been bred in mainland China. Due to a clear genetic relationship and good symmetry, 'ROC1' is proved to be a good parent for crossbreeding (Figure 3-13).

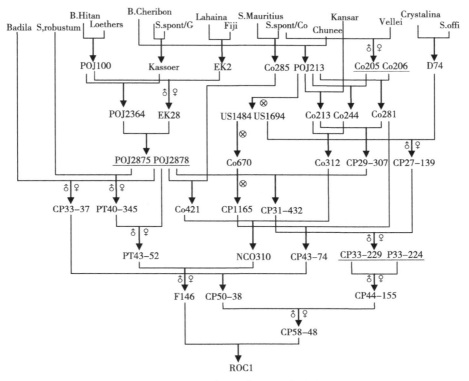

Figure 3-13 The pedigree of ROC1

'CYN73-204': The parental cross is 'CYT57-423' × 'CP49-50'. It was selected as an excellent breeding parent material by the Crop Research Institute, Guangdong Academy of Agricultural Sciences, in the early 1970s. It has the agronomic characters of thick stem diameter, high yield, strong stress resistance, easy flowering, large pollen, and strong vitality. It was a frequently used parent in sugarcane crossbreeding in China since the 1990s, which can be used as both the male and female. Eight offspring varieties have been bred, including 'CYT93-159' and 'CYT00-236'. It is also an essential parent for innovation in China. As shown in Figure 3-14, the consanguinity is relatively straightforward. But there is still some networked kinship. Though the offspring showed excellent agronomic characters, there is a slight breakthrough in this variety's traits.

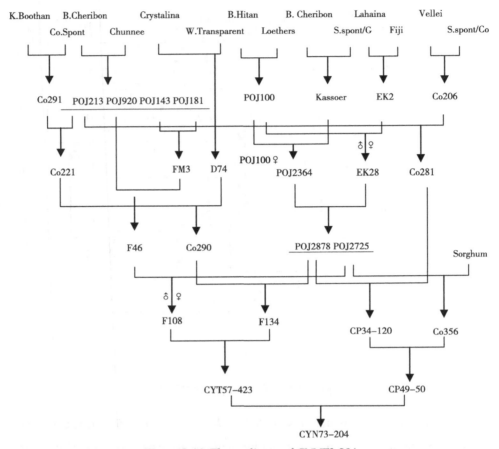

Figure 3-14 The pedigree of CYN73-204

'ROC10': The parental cross of 'ROC10' is 'ROC5' × 'F152'. TWSRI bred it. It is medium maturity, high yield, high sugar content, slow and nice seedling, good ratoonabilty, more millable stalks, thicker stem, resistance to wind and lodging. It is suitable for planting in warm and hot conditions with sufficient water and fertilizer. Since the mid-1990s, seven sugarcane varieties have been bred in China by using 'ROC10' as the parent. The family tree of 'ROC10' is shown in Figure 3-15. The consanguinity is relatively straightforward, but there is still some crossover. It could breed good offspring but with no breakthrough of the traits.

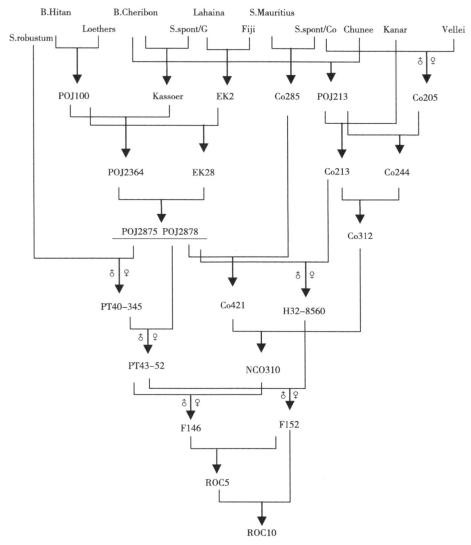

Figure 3-15 The pedigree of ROC10

'ROC22': The parental cross is 'ROC5' × 'PT69-463', developed by TWSRI. The agronomic characters of 'ROC22' are as follows: medium-thick stems, medium maturity, high sugar content, reasonable germination rate, high tillering, the slow growth rate in the early stage, the fast growth rate in the middle and late stage, developed No. 57 hair group, easy to defoliate, resistance to lodging, no flowering, strong ratoonability, and inverted conical internodes. It shows a thick base and small tips. It exhibits wide adaptability and large promotion areas in mainland China. 'ROC22' has made an essential contribution to the development of the Chinese cane sugar industry. Since the early 21st Century, seven sugarcane varieties have been bred in mainland China using 'ROC22' as the parent. It can be seen from pedigree Figure 3-16 that 'ROC22' has a clear blood relationship, but it is still networked to some extent. Though it had bred some varieties, the improvement of its offspring is limited.

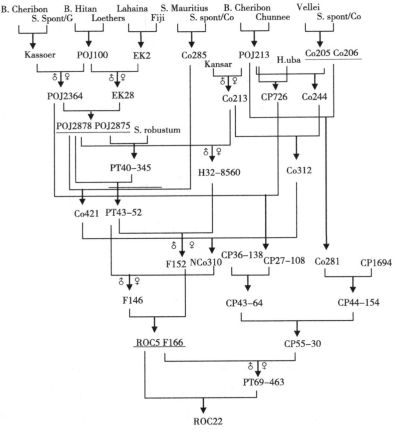

Figure 3-16 The pedigree of ROC22

'ROC25': The parental cross is 'PT79-6048' × 'PT69-463'. TWSRI developed it. It shows medium stem, early mature, high sugar content, orderly germination, fast growth rate, more stalks, good ratoonabilty. There are some promotion areas of 'ROC25' in mainland China. Since it was used as hybrid parents at the end of the 20th Century, the progeny exhibited more stems, strong ratoonability, fast growth, and drought resistance. Five sugarcane varieties have been cultivated with 'ROC25' as a male parent. As shown in Figure 3-17, the genetic relationship of 'ROC25' is clear, but the parents' networking is serious. There will be some defects in the trait in the offspring of 'ROC25'.

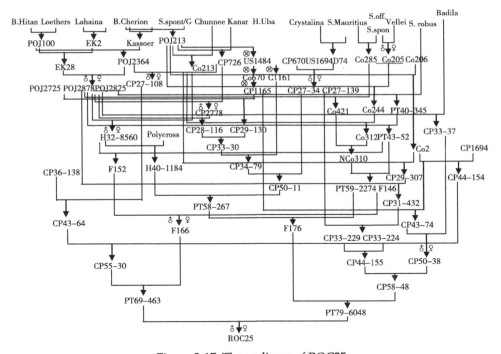

Figure 3-17 The pedigree of ROC25

'Ke5': The parental cross is 'POJ2878' × 'B3412'. It is medium stem diameter, medium late maturity, strong growth potential, medium ratoonabilty, accessible blooming, large pollen, good pollen viability, and cold resistance. Five sugarcane varieties have been cultivated with 'Ke5' as the paternal parent in mainland China since the 1990s. As shown in Figure 3-18, the paternal parent of 'B3412' is unknown. The genetic relationship of 'Ke5' is generally clear and less networked kinship. However, the crossing of 'POJ2878' and 'B3412' is asymmetric, which belong to F_2 and F_3 hybrids, respectively. So it can be used as an excellent parent, but there is some limitation in the traits of the offspring.

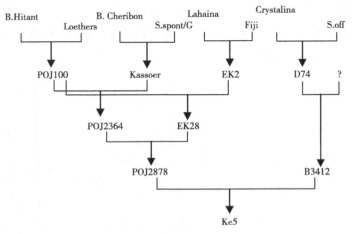

Figure 3-18 The pedigree of Ke5

'CYT85-177': The parental cross is 'CYT57-423' × ('CP57-614' + 'CP72-1312'). A double male parents system bred it in Guangzhou Sugarcane Industry Research Institute (GZSRI). This variety shows medium and late maturity, medium and large stem diameter, high yield, high sugar content, red and cylindrical stalks, large and rhomboid bud with prominent grooves. The leaves are green, erect, and drooping at the tip. The No. 57 hair groups are developed. It grows slowly in the early stage and grows fast in the middle and late stages. It also shows strong ratoonability. It is resistant to Mosaic and Smut disease. The progeny also has good resistance to drought, Mosaic, and Smut. In recent years, five sugarcane varieties have been bred by hybridization utilization of 'CYT85-177'. Because its male parent is unknown, the pedigree chart is omitted.

'CYT91-976': The parental cross is 'CYN73-204' × 'CP67-412'. It was also bred in GZSRI. It has the traits with medium and large stem diameter, high yield and high sugar content, high plant height, vigorous growth in the whole period, fast and regular germination, strong tillering ability, lots of stalks, resistance to Mosaic disease and Smut, good drought and wind resistance and strong ratoonability. Five offspring varieties have been successfully cultivated since 2002. As shown in Figure 3-19, the genetic relationship of 'CYT91-976' is clear but shows serious networking. Like several other top ten parents, there are some defects in the traits of their offspring.

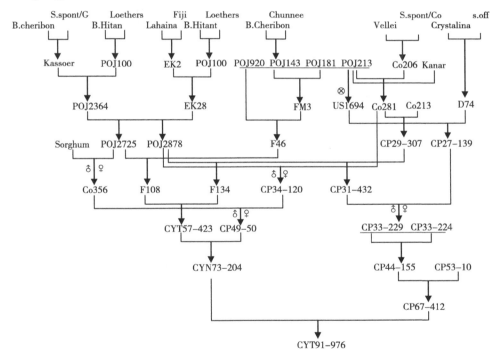

Figure 3-19 The pedigree of CYT91-976

'ROC23': The parental cross is 'F166' × 'PT74-575'. It was bred in TWSRI. This variety exhibits early maturing, high sugar content, medium and large stem diameter, fast and regular germination, fast growth in the early and middle stages, and well growth vigor in the whole period. It shows a thick stem base, easily defoliate, strong lodging resistance, drought resistance, and resistance to Smut. The offspring have a higher sugar content and better ratoonability. Three offspring varieties were successfully cultivated since 2002. As seen in Figure 3-20, the consanguinity relationship of 'ROC23' is the same as several other top ten parents, displaying a clear genetic relationship but serious networking. It can breed suitable varieties, but the offspring still have defects in some traits.

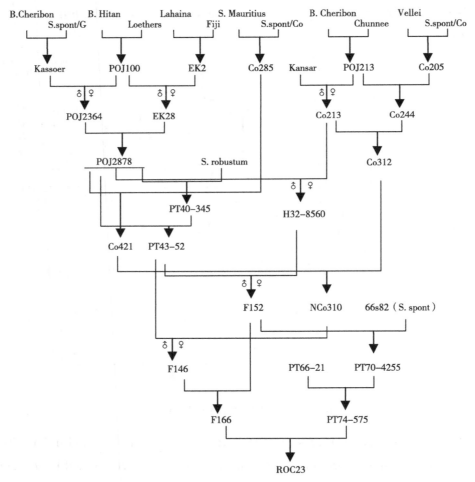

Figure 3-20 The pedigree of ROC23

References

BIAN X, DONG L H, SUN Y F, et al., 2014. Principal component analysis of drought resistance related traits of *Saccharum spontaneum* L. and its F_1 hybrids. Agricultural Research in the Arid Areas, 2014, 32 (3):56-61.

BIAN X, JING Y F, TAO L A, et al., 2015. Difference-similarity analysis of hybrid combinations with wild sugarcane blood in Yunnan. Sugar Crops of China, 37 (2):29-31.

CHEN R K, XU L P, LIN Y Q, et al., 2011. Modern sugarcane genetic breeding. Beijing:China Agriculture Press (in Chinese).

CHU L B, 2000. "YN" Study on sugarcane breeding system-application "Hetero geneous composite separation theory" A super-superior germplasm of Yunnan *Saccharum Spontaneum* L. F_1 with high sugar content was obtained. Sugarcane (4):22-33.

FAN Y H, CHEN H, SHI X W, et al., 2001. RAPD Analysis of *Saccharum spontaneum* from different ecospecific colonies in Yunnan. Acta Botanica Yunnanica, 23 (3):298-308.

HU X, XIA H M, YANG L H, et al., 2021. Identification and evaluation the resistance to smut in sugarcane parental clones. Sugar Crops of China, 43 (1): 45-50.

JING Y F, DONG L H, SHUN Y F, et al., 2013. Genetic analysis of drought resistance of different ecotypes in Yunnan. Journal of Hunan Agricultural University, 39 (1):1-6.

LANG R B, JING Y F, AN R D, et al., 2015. Evaluation by DTOPSIS on cross combinations from Yunrui innovative parent lines and introduced sugarcane lines. Sugar Crops of China. 37 (5):43-45.

LI Y R, 2010. Modern sugarcane science. Beijing:China Agriculture Press (in Chinese).

LIU J Y, CHEN X K, FU J F, et al., 2003. Analysis on the genetic relationship of several fine sugarcane lines[J]. Sugar Crops of China (2):1-5.

LIU X L, SU H S, LIU H B, et al., 2014. Correlation and clustering relationship analysis of Yunnan octoploid clones of *Saccharum spontaneum* in China on the basis of yield and quality related traits. Southwest China Journal of Agricultural Sciences, 27 (4):1382-1386.

LUO J S, 1984. Sugarcane. Guangzhou:Guangdong Sugarcane Society (in Chinese).

MIRJKAR S J, DEVARUMATH R M, NIKAM A A, et al., 2019. Sugarcane (*Saccharum* spp.):breeding and genomics. In:AL-KHAYRI JM. et al. Advances in plant breeding strategies:industrial and food crops. 6. ed. Switzerland:Springer, 11,363-406.

PENG S G, 1990. Sugarcane Breeding. Beijing:China Agriculture Press (in Chinese).

TAO L A, DONG L H, JING Y F, et al., 2014. Compositive evaluating of the grey closeness degree for the hybrid F_1 of the *Saccharum* species. Journal of Plant Genetic Resources, 15 (6):1248-1254.

TAO L A, JING Y F, DONG L H, et al., 2020. Integrated evaluation of the YR15 sugarcane innovative germplasm with different ways of crossing by DTOPSIS. Sugar Crops of China, 42 (1):13-21.

TAO L A, JING Y F, DONG L H, et al., 2011. Genetic analysis of main traits in descendants of crossing with *Saccharum spontaneum* 82-114 in Yunnan, Journal of Plant Genetic Resources, 12 (3):419-424.

TAO L A, YANG L H, AN R D, et al., 2015. Comprehensive analysis and comparison of five methods for 08 series *Saccharum spontaneum* F_2 in Yunnan. Southwest China Journal of Agricultural Sciences, 28 (5):1907-1915.

TAO L A, ZHANG J R, 1997. A preliminary study on the characteristics of S. *spontaneum* L. and E. *arundinaceum* F_1-generation superior materials with resistance to chopping and sugar transformation. Sugarcane (2):9-11.

TIAN C Y, TAO L A, YU H X, et al, 2017. Drought resistance evolution of F_1 and F_2 hybrids from five climatic ecotypes *Saccharum spontaneum* L. Agricultural

Sciences in China, 50 (22):4408-4421.

WU C W, 2002. Analysis of the utilization and efficiency of the parents for sugarcane sexual hybridization in Yunnan. Sugarcane and Canesugar (4):1-5.

WU C W, 2005. Discussion on Germplasm Innovation and Breeding Breakthrough Varieties in Sugarcane. Southwest China Journal of Agricultural Sciences, 17 (6):858-861.

WU C W, 2018. Discussion on breakthrough of inbreeding and interleaved use of sugarcane parents in creating new sugarcane parents. Sugar Crops Improvement, Biotechnology, Bio-Refinery, and Diversification:Impacts on Bio-based Economy, Udon Thani, Thailand, 407-410.

WU C W, 2021. Advances in Breeding New Independent Parents of Sugarcane by the Peer to Peer Hybridization in Yunnan. Sugar Crops of China, 43 (3):37-41.

WU C W, ZHAO J, LIU J Y, et al., 2014a. Modern sugarcane seed industry. Beijing:China Agriculture Press (in Chinese).

WU C W, ZHAO P F, XIA H M, et al., 2014b. Modern cross breeding and selection techniques in sugarcane. BeiJing:Science Press (in Chinese).

YU H X, TAO L A, TIAN C Y, et al., 2019a. Utilization of *S. robustum* 57NG208 in breeding of YR-series parental clones, Sugar Crops of China, 41 (2):1-7.

YU H X, TIAN C Y, JING Y F, 2019b. Principal component clustering analysis and evaluation of F_2 generation from Yunnan *Saccharum spontaneum* L. innovation germplasm. Journal of Plant Genetic Resources, 20 (3):624-633.

ZHANG M Q, WANG H Z, BAI C, et al., 2006. Genetic improvement and efficient breeding of sugar crops. Beijing:China Agriculture Press (in Chinese).

4 Disease Resistance Breeding and Its Effect in Sugarcane

4.1 Important Sugarcane Diseases and Their Harm

The disease is an essential factor affecting sugarcane production. At present, there are 130 known sugarcane diseases in the world. By investigating sugarcane planting areas in mainland China in the 1990s, 39 sugarcane diseases were found, including 23 fungal diseases, 3 bacterial diseases, 2 virus diseases, 8 sugarcane nematodiasis diseases, and 1 parasitic plant disease. The infectious diseases include 29 fungal diseases, 5 bacterial diseases, 8 virus and phytoplasma diseases, 8 nematode diseases, 2 parasitic plants, and the rest are non-infectious diseases (Li, 2010). There are various climatic types in sugarcane growing areas in China. Different diseases are closely related to cultivated varieties in different ecological types of sugarcane areas and then to the climatic conditions and cultivation systems in different regions. Therefore, the occurrence and spread of various sugarcane diseases, disease types, and distribution are not the same. The diseases that greatly influence sugarcane production in China are mosaic disease, yellow leaf syndrome, smut, rust, Pokkah Boeng, ratoon stunting disease (RSD), pineapple disease, yellow spot, brown stripe disease, and nematode disease.

The harm of important sugarcane diseases to sugarcane production in China's sugarcane growing areas (except Taiwan Province) is as follows: in the 1950s and 1960s, except for pineapple disease, other diseases only occurred sporadically, and no disease was found to be prevalent in sugarcane areas. However, since the 1970s, sugarcane diseases have been expanding in sugarcane areas. Some sugarcane diseases are often prevalent in some areas, which brings great losses to sugarcane production. For example, yellow spot disease was prevalent in Taishan, Haifeng, and Haikang counties of Guangdong Province in 1973. Only three heart leaves of the infected sugarcane plants were green, and the rest of the leaves were all dried up, resulting in a massive loss of sugarcane yield and sugar content. In 1972 and 1977, smut was prevalent in Xuwen, Guangdong, Neijiang, Sichuan, and Hunan, and the incidence rate was about 30%. The incidence rate of serious fields was more than 90%, almost to the extent of the destruction. In 1981, eyespot disease was prevalent in Wenshan County, Yunnan Province, with an area of 91 hm^2 and a loss of 600–700 tons of sugarcane. From 1978 to 1979, the Pearl River Delta in Guangdong Province broke out with an area of over 1333 hm^2. In 1979, sugarcane mosaic disease was prevalent in Jinhua and other counties of Zhejiang. The incidence rate of infected sugarcane fields was as high as 70% to 80%. In recent years, it was found that mosaic disease occurred more commonly in sugarcane areas of Guangdong, Fujian, Sichuan, and Jiangxi provinces. In the past, sugarcane rust only occurred in Taiwan. Fujian Province first reported the occurrence of rust in 1983. Later, it was found in the Zhongshan and Panyu areas of Guangdong Province. In recent years, some sugarcane areas in Yunnan Province also suffered from rust. It shows that rust began to spread from Taiwan Province to the mainland sugarcane areas and occurred in many sugarcane areas in the mainland. In 1989, sugarcane shoot rot occurred seriously in Dongguan and other sugarcane areas of Guangdong Province, with an area of more than 867 hm^2.

Sugarcane ratoon stunting disease, smut, and mosaic disease are prevalent in the Yunnan sugarcane area are seriously endangered. The detection rate shows that the incidence rate of ratoon stunting disease is 70%, 100% and 10%–30% for drought, and 60% for drought and water.

It can be seen from the above situation that sugarcane diseases have seriously threatened sugarcane production in China. Since the 1950s, with the development of the chemical industry, various chemicals have been widely used in production. Although the harm of sugarcane diseases is reduced, scientists have noticed that the long-term use of chemicals increases production costs, causes environmental pollution, and damages the ecosystem. Therefore, the integrated control system of sugarcane diseases has been gradually developed. The methods of disease-resistant varieties, cultivation control, biological control, and chemical control are applied comprehensively to keep the pathogen below the economic threshold. The practice has proved that the most economical and effective measure is to select new varieties of resistant sugarcane. The selection of disease-resistant varieties is an essential basis for the establishment of an integrated control system. It can inhibit the number of pathogens, reduce the harm of sugarcane disease, improve the control effect, and reduce the environmental pollution caused by the abuse of pesticides and maintain ecological balance. It plays a critical role in the sustainable development of the sugarcane industry and the safety of sugarcane products in China. Therefore, disease resistance breeding is becoming more and more essential, and it's imperative.

4.2 Disease Resistance Breeding and Strategy in Sugarcane

Sugarcane disease is one of the biggest threats to sugarcane production, which is recognized by the primary sugarcane-producing countries in the world. Sugarcane disease is also one of the crucial reasons for variety degradation. Breeding disease-resistant varieties are an essential means to resist sugarcane diseases. The threat of sugarcane diseases to sugarcane production is the main driving force of variety replacement. Sugarcane disease-resistant varieties have made significant contributions to the development of the cane sugar industry in the world. For example, from 1885 to 1896, Java, Indonesia, suffered from severe sugarcane atrophy, and the local cultivar 'B.Cheribon' suffered from severe disease and suffered heavy losses in sugarcane production. The whole sugar industry in Java was on the verge of collapse. A series of disease-resistant varieties, such as 'POJ2878', were bred to control the cane sugar industry atrophy. In the late 1920s, due to the long-term cultivation of 'D74', Otaheite, Creole, and Louisiana red, the varieties became single, causing the severe occurrence of several diseases such as mosaic disease, root rot, and red rot, which almost destroyed the whole Louisiana sugar industry. Later, the introduction of disease-resistant varieties 'POJ213' and 'POJ234' saved the Louisiana cane sugar industry. The mosaic disease occurred in Taiwan Province in 1916. It was forced to destroy all the susceptible varieties POJ161 from 1919 to 1925 and replant resistant varieties such as 'POJ2878' 'POJ2725' and 'POJ2883'. The disease occurred again in 1947, and 'F108', which had been planted in large areas, had to be replaced by the resistant variety 'NCo310', and the crisis was over again (Zhang, 1998). In Argentina, 80 % of sugarcane was destroyed due to the outbreak of smut in the mid-1930s. Finally, the sugarcane industry was saved through variety updating (Que et al., 2009). In the past 60 years since the 1950s, 'zhucane' 'F134' and 'CGT11', the largest sugarcane varieties in China's mainland, were seriously infected by smut and were eliminated successively.

The sugarcane diseases in the above countries (regions) have caused considerable losses to the sugarcane industry. A series of disease-resistant varieties have been bred through disease resistance breeding, which has saved the local sugarcane industry. Breeding disease-resistant varieties are the most economical and effective method to solve most of the sugarcane diseases. The major sugarcane planting countries in the world attach great importance to the disease resistance breeding of sugarcane, take the disease resistance of varieties as a vital index of sugarcane breeding, and breed many excellent varieties with disease resistance safety sugarcane production.

Although disease-resistant varieties are the most economical and effective way to solve most sugarcane diseases, the cost-effectiveness ratio should be considered in disease resistance breeding. If other economic methods can control some diseases, it is not necessary to emphasize disease resistance breeding. Not every disease-resistant variety should start from parent selection, but different disease resistance breeding should be adopted according to different diseases strategy.

4.2.1 Strategy of disease resistance breeding

The strategy of disease resistance breeding widely used in the world's major sugarcane planting countries can be roughly divided into the following three methods.

4.2.1.1 Diseases without emphasis on disease resistance breeding

The purpose of disease resistance breeding for most diseases (in fact, almost all diseases) is to maintain yield. Some diseases do not cause loss of sugarcane yield (or a small amount of yield reduction), so disease resistance breeding is not emphasized. For diseases that can be controlled by physical or chemical methods (such as seed treatment, cultivation measures, etc.), such prevention and control must be adequate, economical, or have little impact on the environment breeding for disease resistance. For example, sugarcane pineapple disease can be effectively controlled as long as the seedlings are disinfected with fungicides before planting. For example, in sugarcane ratoon stunting disease, through hot water treatment or tissue culture detoxification of seedlings, disinfection of harvesting tools, and propagation in the nursery, the nursery provides disease-free seedlings for field production for production to effectively reduce the loss of disease on production and achieve good control effect. As long as there is no pathogen, the disease will not occur. In this case, both susceptible and resistant varieties can be planted. Therefore, the disease resistance of these diseases can be less considered in the breeding plan. In addition, the cultivation measures can prevent and control sugarcane mosaic disease for the spread of sugarcane mosaic disease with many physiological races. It is possible to set up a breeding base for healthy seedlings in more minor and relatively isolated areas of aphids and provide disease-free seedlings for production. This measure can play a good role in the prevention and treatment of mosaic disease.

4.2.1.2 It is not necessary to consider the disease resistance of parents in the selection of parents, but it is necessary to select the disease resistance of their offspring

Some sugarcane varieties are resistant to some diseases, such as the sugarcane Pokkah Boeng disease in the sugarcane field. For example, suppose a clone is susceptible to Pokkah Boeng. In that case, it will develop at all stages of the breeding process and under high temperature and humidity conditions, and it will be eliminated artificially. The excellent clones left behind are disease-resistant clones. Therefore, when selecting parents, breeders need not consider the resistance of hybrid parents (Li et al., 1999).

4.2.1.3 Diseases that emphasize disease resistance breeding

It can cause serious harm to sugarcane production. The effect of physical or chemical control is not good, or some diseases which are not economical to control by physical or chemical methods, such as sugarcane smut disease, etc., or a minor sugarcane disease rises to significant disease in sugarcane area, it is necessary to adopt disease resistance breeding. Disease resistance breeding should aim at the primary and secondary damage degree of pathogens of various local diseases and determine the resistance index of disease resistance breeding. Still, the index should not be too many. Otherwise, the selection intensity of index characters will be reduced. The resistant parents must be selected, and their offspring should be identified for disease resistance. In the breeding program, breeders should select at least one parent that is resistant to disease, and at the early stage of the breeding program, the clones susceptible to natural conditions should be eliminated in the early stage of the breeding program and in the middle and late stage, the disease resistance of hybrid progenies that are expected to become new varieties will be identified by artificial vaccination and natural infection.

4.2.1.4 Rational application of disease resistance breeding

The effect of disease resistance breeding needs to pay attention to the following aspects in sugarcane production.

(1) Reasonable distribution of sugarcane varieties

No disease-resistant variety can resist all diseases, and a perfect variety is not in line with the objective reality. If the variety performs well, it will be promoted vigorously without considering the variety layout, resulting in serious simplification of sugarcane varieties, threatening the sustainable, stable, and healthy development of sugarcane production. Sooner or later, some minor sugarcane diseases or physiological races will accumulate for a long time; it has been proved by the history of sugarcane planting in the world that the disease-resistant varieties become susceptible varieties that need to be replaced.

(2) Do an excellent job in quarantine work

If there is no strict introduction and quarantine system, it is bound to cause diseases. When a certain number of introduced diseases accumulate, they will occur seriously under one condition, resulting in heavy losses of sugarcane production. According to the investigation, there were only 14 sugarcane diseases in the Guangdong sugarcane region in the 1950s, but 35 diseases were found in the 1980s, 21 more than in the 1950s, especially 12 fungal diseases caused the severe threat to sugarcane production. Especially in recent years, many sugarcane varieties have been introduced from abroad and transferred to each other. Although it has a significant role in accelerating the improvement of sugarcane varieties, the strict quarantine system of diseases and insect pests has not been implemented in the link of introduction and seed transfer, which has brought severe hidden dangers to the safe production of sugarcane and put forward new topics for disease resistance breeding.

The introduction and quarantine system must be strengthened in sugarcane interplanting. Firstly, the disease status of the introduced sugarcane area should be mastered, and the introduction from the disease area should be avoided as far as possible, and the field without disease should be selected for introduction from the disease area; secondly, the introduced sugarcane species should be propagated intensively, and the disease monitoring should be strengthened. Once the disease is found, it should be prevented and destroyed in time to control its spread.

(3) The prediction of possible diseases and the monitoring of disease status should be made well

To do an excellent job predicting sugarcane diseases, many countries and regions have conducted cooperative research on identifying disease resistance. The sugarcane production and promotion varieties were sent to the cooperative countries or regions to identify the resistance of primary diseases under different conditions and exchange the research results. It will enable the cooperative organizations to know the resistance degree of the varieties produced and popularized in the sugarcane area to the related diseases. Therefore, it will be of great help to formulate variety management and breeding plan to avoid the promotion of susceptible varieties or the use of susceptible parents in the breeding plan to improve the disease resistance of varieties. At the same time, the development and stability of primary and secondary diseases should be monitored. The disease can be controlled in a stable state through the use and reasonable distribution of resistant varieties, forming an excellent system to achieve the long-term and stable disease resistance of the series of resistant varieties. However, the stability is relative. Host, pathogen, and environmental conditions often lead to the sudden disappearance of disease resistance of varieties in a specific sugarcane region, and the disease resistance becomes susceptible. If the environment is abnormal and suitable for disease occurrence, it will lead to a disease epidemic. Usually, when the environment is normal, the disease is in a latent or local occurrence. In addition, the resistance degree of a variety will gradually decrease after breeding in an environment without disease pressure for many generations. However, the loss of resistance is a slow, gradual process.

The monitoring of disease status is helpful to actively grasp the process of disease control from stable to unstable transformation to adopt variety rotation, promotes multiple varieties in the same sugarcane area, and timely promotes new disease-resistant varieties to achieve the effect of relatively stable overall disease resistance.

4.2.2 Genetic basis of disease resistance

The genetic background of sugarcane is the basis of disease resistance breeding. The disease resistance of sugarcane can be inherited, and genes control the disease resistance. According to the response of varieties to different physiological races of pathogens, disease resistance can be divided into specific and non-specific resistance. The advantage of specific resistance is that the resistance is generally strong, but the disadvantage is that it often loses its resistance due to the changes of physiological races of the disease; on the contrary, non-specific resistance is lower than that of specific resistance, but it is relatively stable and not easy to lose. The inheritance mode of disease resistance genes includes dominant and recessive inheritance; the number of disease-resistant genes controlling diseases includes single gene and multi-gene control diseases (Li et al., 1999). It is generally believed that plant-specific resistance is inherited by quality traits controlled by a single gene, while quantitative traits controlled by multiple genes inherit nonspecific resistance. At present, it is considered that the resistance of many sugarcane diseases is a quantitative trait, which is controlled by multiple genes with additive effects. It is necessary to know the heritability of disease-resistant traits when selecting parent combinations. Some research results show that the narrow-sense heritability of disease-resistant traits is 0.2−0.7. Parents should generally choose the varieties or lines with good economic traits to cross with the resistant parents. If the disease resistance of the resistant parents belongs to single gene inheritance, another parent with good economic traits can be used as the recurrent parent for "backcross" to transfer the disease resistance traits to the recurrent parents; if the disease-resistant parents have more excellent traits, only single cross with the parents with good economic traits can be carried out. If polygenes control the disease resistance of the parents, or the disease has different physiological races, it is necessary to carry out multiple crosses" to transfer multiple resistance genes to a new variety, to improve the disease resistance of varieties.

4.2.3 Breeding of disease-resistant varieties

4.2.3.1 Collect the source of resistance

Germplasm resources are the material basis of crop breeding. The introduction, collection, evaluation, and utilization of sugarcane germplasm resources is a meaningful way to broaden the genetic basis and improve breeding efficiency. It is necessary to strengthen the identification and evaluation of sugarcane germplasm resources for disease resistance and construct the database of sugarcane disease resistance resources. Abundant disease-resistant germplasm resources are the material basis for breeding disease-resistant varieties. Only by mastering abundant disease-resistant germplasm resources and cultivating breakthrough resistant sugarcane varieties can it be realized. The disease-resistant sugarcane germplasm resources include locally cultivated varieties, hybrid parent materials, introduced varieties or materials, wild sugarcane species, and related plants. Understanding the disease resistance of disease-resistant germplasm resources to the primary diseases in this region is the basis for breeding disease-resistant varieties.

4.2.3.2 Identification of disease resistance

With the development of artificial inoculation technology, new progress has been made in disease resistance breeding. The more research on physiological races of disease is, the more targeted the disease resistance breeding is. The identification of sugarcane disease resistance is generally divided into early identification and late identification. In the early stage, the seedlings were identified by artificial inoculation in a greenhouse from seed germination to planting, and the performance of disease resistance was mainly observed; in the later stage, the best materials were selected for artificial inoculation identification in the field, and the agronomic characteristics were observed. The inoculation technology varies with the type of pathogen. If it is the inoculation test of acute and epidemic diseases, it should be far from the raw sugarcane area to prevent the spread of the disease. A unified identification standard should be established to exchange and compare the information on disease resistance in different sugarcane regions in China (or abroad).

The research of disease resistance identification and identification technology includes: Firstly, objective to understand the correlation between artificial vaccination and natural infection, plot (or laboratory) test, and field infection. Secondly, objective to collect and understand the pathogenic bacteria and their physiological races needed for vaccination. Thirdly, determine the inoculation method, concentration, and time used in the test. Fourthly, determine the number of test repetitions and the specific test design scheme.

4.2.3.3 Cross utilization of disease-resistant germplasm resources and breeding of disease-resistant varieties

Sexual cross-utilization of sugarcane disease-resistant germplasm resources is the primary way of disease resistance breeding. It is a method of breeding new sugarcane disease-resistant varieties by crossing parents and creating new varieties. So far, the vast majority of sugarcane disease-resistant varieties are obtained through sexual cross-breeding. Sugarcane is an aneuploid allopolyploid crop with a complex genetic basis. The progenies of a cross between varieties have extensive segregation and many variations and have a great chance to produce good individuals with disease resistance. Sugarcane disease resistance breeding is usually combined with conventional sexual cross-breeding. At least one of the two parents is a resistant parent. In selecting hybrid progenies and excellent agronomic traits, one or more disease resistance indexes should be selected.

4.2.4 A new way of sugarcane breeding for disease resistance

In recent years, the rapid development of molecular technology provides a new way for disease resistance breeding. Molecular technology of disease resistance breeding is an interdisciplinary subject of pathology, breeding, and molecular technology. Its application in disease resistance breeding mainly includes transgenic disease resistance breeding and molecular marker-assisted disease resistance breeding.

4.2.4.1 Transgenic disease resistance breeding

According to disease resistance breeding, the target genes were isolated from donor organisms and introduced into target varieties by particle bombardment or Agrobacterium-mediated method. After the screening, genetic engineering bodies' stable expression was obtained, and the resistant transgenic sugarcane seedlings were obtained. The transgenic sugarcane-resistant new varieties were bred by disease resistance verification tests and field selection, which could resist the invasion of target diseases. The procedure of transgenic disease resistance breeding is as follows: Firstly, screening and isolation of resistance gene and construction of its vector. Secondly, recombined in vitro. Thirdly, gene gun bombardment or agrobacterium-mediated. Fourthly, the transformants were obtained and combined with conventional breeding; the disease resistance and genetic stability were evaluated by disease resistance verification experiment and field screening. Fifthly, new transgenic sugarcane varieties with disease resistance were obtained, and their safety was evaluated. Sixthly, market development and application.

Compared with conventional breeding, transgenic disease resistance breeding has four advantages: Firstly, it broadens the available disease resistance gene resources. Secondly, it provides a new way for breeding new varieties with high quality, high yield, and disease resistance. Thirdly, it can carry out directional variation and directional selection for the target traits of crops. Fourthly, it can significantly improve the selection efficiency and speed up the breeding process. Currently, sugarcane virus resistance transgenic research mainly focuses on Fijian sugarcane disease (FDV) and sugarcane mosaic disease (SCMV). Mcqualter et al. (2004) transferred the genome fragment of sugarcane Fijian disease virus controlled by maize Ubi promoter into sugarcane 'Q124' and obtained a transgenic sugarcane plant resistance Fijian disease virus. Ingelbrecht et al. (1999) introduced the coat protein gene (CP) of Sorghum Mosaic Virus SCH strain into sugarcane and obtained the virus-resistant transgenic plants (Gan et al., 2013).

4.2.4.2 Molecular marker-assisted breeding

Molecular markers based on DNA polymorphism have been widely used in crop genetic map construction, breeders have paid marker location of critical agronomic traits genes, genetic diversity analysis of germplasm resources, variety fingerprint, purity identification, etc., molecular marker-assisted selection (MAS) more attention. MAS breeding can be used for large-scale early selection of sugarcane hybrid progenies. Disease-resistant materials can be selected by screening molecular markers closely linked to disease resistance genes. Compared with wheat, rice, maize, and other crops, sugarcane molecular markers development is slow, the construction of molecular map and the research of disease resistance gene linkage markers are relatively backward, mainly due to the large genome of sugarcane (typical about 3,000 Mb), and the genetic background is complex. Nevertheless, the research on sugarcane molecular markers has made gratifying progresses. More than ten molecular maps of sugarcane have been constructed, and several DNA molecular markers linked to disease resistance genes have been obtained. Some reports on genetic diversity, phylogeny, and genetic relationships among sugarcane varieties using molecular markers (Que et al., 2005).

In applying molecular markers in sugarcane breeding for disease resistance, firstly it is necessary to find the markers stably linked to the target genes; secondly, the detection methods must be simple, rapid, reproducible, and economical. Because it is challenging to obtain near-isogenic lines (NIL) and DH populations in sugarcane, bulked sergeants analysis (BSA) is easy. At present, it occupies a critical position in the identification of linkage markers. BSA method can analyze the segregation population of any cross progeny of resistant and susceptible parents. By identifying the disease resistance of individuals in the population, several extreme populations were selected. Resistant and susceptible gene pools were constructed to detect markers linked to targeting traits. DNA markers (such as RFLP, AFLP, RAPD, SSR, SNP, etc.) closely linked to disease resistance genes were screened by the BSA method and then cloned and transformed into scar markers or STS markers to further improve the stability of marker analysis and the ability of large-scale marker analysis.

4.3 Effectiveness and Progress of Disease Resistance Breeding

4.3.1 Breeding of sugarcane varieties resistant to smut

The Sugarcane Smut (*Ustilago scitaminea* Syd.) has become one of the primary diseases in the world. Sugarcane Smut can cause severe loss of cane yield and sugar content of susceptible varieties, and some high-yield and high sugar varieties such as 'NCo310' 'F134' and 'CP73-351' have been eliminated. Therefore, several major sugarcane producing countries in the world, such as the United States, Cuba, India, and Brazil, have taken the resistance of sugarcane clones to smut as a primary target of variety selection, the susceptible plants were eliminated by natural infection, and the sensitive materials were further eliminated by artificial injection in the middle and late stage of breeding. In Louisiana, from 1982 to 1983, 75% of the candidate clones were moderately susceptible to smut, and more than 50 % of the candidate clones were eliminated. Sugarcane Smut is also the most common and severe disease in sugarcane production in China. At present, 'ROC16' and 'ROC22', with the largest planting area in China, have also been seriously infected by smut in many sugarcane areas and face the situation of being replaced by new resistant varieties.

4.3.1.1 Resistance mechanism of sugarcane to smut pathogen

The pathogen of sugarcane smut mainly invades the sugarcane bud through the scale crevice of lateral sugarcane bud. With the growth of sugarcane, the mycelium also grows upward with the growth point. After a certain period of sugarcane growth, the pathogen is stimulated to form a smut whip, thus harming the sugarcane. There are two different mechanisms of sugarcane varieties' resistance to smut. One was morphological resistance, mainly due to the physical structure and performance of the sugarcane bud scale that the pathogen cannot invade sugarcane bud. Padmanaban et al. (1988), Gloria et al. (1999), and Gong et al. (1996) studied the relationship between the structural characteristics of buds and the disease resistance of sugarcane varieties. The results showed that the bud type varieties with large bud, deep bud groove, no bud wing, and germinal pore were more susceptible, while the subapical sprout type with no major bud groove, compact bud scale, and not at the top showed robust resistance. The other was physiological and biochemical resistance, which showed that the mycelium could not spread in the stem tissue and could not further infect sugarcane buds. Previous studies have confirmed the existence of physiological and biochemical resistance in sugarcane bud scales from the experiments that glycosides in bud scales can inhibit spore germination, and the relationship between the contents of n-dihydroxy phenol, total sugar, free amino acids and glycosides and disease resistance (Chen et al., 2003).

4.3.1.2 Inheritance of sugarcane resistance to smut

The resistance source of sugarcane to smut was from *S. officinarum* and *S. spontaneum,* and the susceptible source was from *S. barberi* and *S. spontaneum* / Co. The genetic characteristics of smut resistance in sugarcane were studied by Wu et al. of Hawaii (1983; 1977), Walker of Barbados (1980), Chao et al. of Louisiana (1990), and Chinese scholars. The results showed that the resistance of sugarcane to smut was moderate or highly heritable, and the resistance of sugarcane to smut was quantitative genetic behavior. It indicated that the improvement of sugarcane resistance to smut could be obtained by crossing resistant parents.

The selection efficiency of sugarcane clones depends on the repeatability of traits crossing the selection stage in time and space. Repeatability is an index to measure the correlation degree of a quantitative trait among multiple individual measurements. Analyzing the repeatability of sugarcane resistance to smut in different generations and environments is helpful to evaluate its resistance level accurately. Chao et al. (1990) showed that the repeatability of new planting sugarcane and first season ratoon sugarcane resistance was 0.60 and 0.75, respectively. The resistance level was significantly correlated with crop season and year. The study of Lin et al. (1996) showed that the resistance repeatability of resistant and susceptible materials was moderate (0.60 for new planting and ratoon crop) and low for medium susceptible sugarcane.

4.3.1.3 Physiological race of sugarcane smut

Two races A and B were found in China, but two were found in China and Brazil. Race 1 tended to be the dominant race (Xu et al., 2000).

The identification results are difficult to compare due to the different hosts used to identify physiological races of sugarcane smut in the world and race naming. DNA fingerprinting technology has been widely used to identify pathogens and races with the development of molecular biology. Some scholars used PCR to detect smut in sugarcane and it can be used to distinguish two mating types (a and b).

4.3.1.4 Breeding of smut resistant varieties

In selecting parents, parents with disease resistance and other good traits should be selected as far as possible.

In the middle and late stages of the breeding program, the resistance of new lines to smut was evaluated by artificial vaccination and natural infection. Because the resistance repeatability of resistant and susceptible materials is moderate, and the resistance performance of clones is affected by genotype and genotype-environment interaction, it needs many years and multi-point resistance identification to evaluate the resistance of varieties accurately.

4.3.1.5 Identification method of sugarcane resistance to smut

Resistance identification can be divided into artificial inoculation identification and natural infection identification.

(1) Artificial inoculation

According to the infection way of smut and the resistance mechanism of sugarcane varieties to smut, the primary resistance identification of sugarcane varieties was artificial inoculation.

A. Inoculation method

Firstly, the spores of smut were collected, and the germination rate was determined. Secondly, the spore suspension containing 5×10^6 spores per milliliter was prepared, and the spore viability was above 90%. 0.02% dilute hydrochloric acid can be added to the mix to destroy the surface ion exchange membrane. Thirdly, the identification materials were inoculated by the dipping method. The inoculated materials without diseases and insect pests were selected. Each material had 30 single buds. The materials were immersed in the prepared spore suspension for 10 minutes. After taking out the materials, they were put into a plastic bag, and then they were moistened at 25 °C for 24 h before sowing. Two replicates were set.

B. Investigation record items

Inoculation date, emergence number, the initial date of smut flagella, accumulated number of clusters, and duration of cluster disease.

(2) Natural infection

In addition to artificial inoculation tests, a natural infection test must be carried out to evaluate the resistance of sugarcane clones to smut accurately.

Firstly, the strain should be planted in the sugarcane area with the most severe smut occurrence under natural conditions to understand the incidence of the strain under such environmental conditions. If only a small number of black whips are produced occasionally, then the strain can be popularized and planted after it becomes a variety. If many black whips are produced continuously during the growth period, they should be eliminated.

Secondly, the intermediate type (moderately resistant or moderately susceptible) clones were planted in sugarcane areas with mild smut. Suppose the incidence rate of intermediate type sugarcane is lower or similar to that of newly planted sugarcane. In that case, the strain can still be planted in a light disease area after becoming a variety. On the contrary, the strain should be eliminated.

4.3.1.6 Classification standard

At present, the main sugarcane planting countries and regions worldwide have a set of evaluation and grading standards for sugarcane resistance to smut. India, South Africa, and Taiwan of China adopt 0-9 classification, Hawaii, Australia, Philippines, and New Zealand adopt 1-9 classification, Cuba adopts 1-5 classification, Mexico adopts 4-level classification (R, M, S, HS), China adopts 1-9 classification. Except for 0-9 grade, 0 is immunity, the essence of other classifications is the same. The results showed that the resistant reaction types were all classified into five types: HR (high resistance), R (resistance), M (intermediate type, some were subdivided into MR and MS), S (susceptibility), and HS (high susceptibility) (Xu et al., 2000).

The other method is to adopt the standard variety method, including a set of known resistance varieties in the tested varieties as the standard control varieties, and compare the resistance performance of the clones to be tested to obtain the relative resistance. The standard variety method was not used in the evaluation of resistance in mainland China. The Sugarcane Research Institute of Yunnan Academy of Agricultural Sciences (YSRI) conducted cooperative research on smut resistance with Australia through "Sino Australian international cooperation". In the experiment, Australian standard varieties were used to evaluate the resistance of identified varieties to smut (Xia et al., 2009). The advantage of the standard variety method is that it can reduce the differences between different locations, crop seasons, and years caused by environmental interaction. However, the standard varieties lack self-bred varieties in China, which need further supplementation and improvement.

Some scholars used tissue staining after inoculation of seed stem to evaluate the resistance of clones quickly. The results showed that there was no significant difference in t-test results between the two methods. Because there were different physiological races of smut, the conclusion of which race should be used could only be expressed the resistance level to this race is shown.

4.3.1.7 Achievements and progress of smut resistance breeding

The resistance of sugarcane varieties to smut has been regarded as one of the indexes for variety evaluation since the state purchased sugarcane varieties which had been tackling critical problems in the Ninth Five-Year Plan and national sugarcane variety examination and approval. China's major sugarcane research institutes pay more attention to breeding smut-resistant varieties and identifying and evaluating germplasm resources. The varieties and clones with resistance to smut have been preliminarily screened as follows: 'CFN83-36' 'CFN91-21' 'CGT84-332' 'CGT90-95' 'CGT94-116' 'CGT94-10' 'CYT85-19' 'CYT85-177' 'CYT85-5' 'CYT85-1722'

'CYT89-240' 'CYT91-976' 'CYT89-526' 'CYZ81-173' 'CCN78-111' 'CYC79-277' 'CYC82-110' 'CYC87-28' and 'Uba' (*S.Sinense* Roxb.) were identified. The varieties and Germplasm with resistance to smut introduced from abroad were preliminarily screened: 'Q171' 'CP89-1509' 'CP85-1308' 'FR93-344' 'Hocp91-555' 'Muck che' 'Mex73-206' and 'Kara karawa' (*S. officinarum*), etc. (Xia et al., 2007; Xia et al., 2009; Huang et al., 2001).

Research is being carried out in China to construct a resistant pool to conduct molecular marker-assisted breeding of sugarcane smut resistance genes with the development of molecular biology. The markers linked to the resistance genes were screened. The resistance differential expression genes were cloned, laying the foundation for applying molecular biotechnology in sugarcane breeding for smut resistance. Xu et al. (2001) obtained molecular markers linked to smut resistance gene by BSA method; Gu et al. (2008) introduced chitinase and β-1, 3-glucanase genes into sugarcane varieties 'ROC10' and 'ROC22', which have high sugar content, it was effectively expressed in the intercellular space, to prevent the invasion of sugarcane smut pathogen and make sugarcane varieties obtain resistance; Orlando et al. (2005) studied the molecular response of sugarcane and smut pathogen after the interaction by cDNA AFLP technology, and obtained 62 differentially expressed genes, 52 of which were up-regulated and 10 were down-regulated; Que et al. (2008) cloned NBS-LRR resistance of sugarcane Kong (2012) will have *KP*4 with high resistance to smut in maize and wheat The gene was transformed into sugarcane cultivar 'ROC22' by Agrobacterium-mediated, and 56 positive transgenic plants were obtained. The crude protein of leaves of the plants had an apparent inhibitory effect on the spore germination of smut pathogen (Gan et al., 2013).

4.3.2 Breeding of sugarcane varieties resistant to mosaic disease

Sugarcane mosaic disease is an essential worldwide sugarcane disease, which is a viral disease. Musschenbrock first described the disease in Java in 1892, and then it was seriously prevalent in Argentina, Puerto Rico, the United States,

Cuba, and other countries and regions. So far, it is still prevalent in the main sugarcane areas of the world. In mainland China, the reasons for the diversification and complexity of the ecological environment, the stereoscopic climate, and the cropping system in the sugarcane area have been introduced frequently in recent years, including a frequent and large-scale introduction from abroad, sugarcane inter planting interval, and long sugarcane growth cycle, long-term continuous cropping, perennial root cultivation, asexual propagation, continuous planting and planting diversification. The sugarcane mosaic disease has become one of the most common and severe diseases in China's Guangxi, Yunnan, Guangdong, Sichuan, and other major sugarcane-producing areas. According to the investigation, the incidence of sugarcane mosaic disease in South China, especially in Guangxi dryland, can reach more than 30%, yield loss is 3%–50%, internode length of the diseased plant becomes shorter, and quality becomes worse, which causes severe economic loss every year. The results showed that the germination rate of sugarcane decreased, the chlorophyll was destroyed, the growth was poor, the tillers were less, and the juice amount was reduced. Generally, the sugarcane yield was reduced by 10 % to 40 %, the reducing sugar in the juice increased, and the crystallization rate of sucrose decreased.

4.3.2.1 Pathogeny

The pathogen of sugarcane mosaic disease is complex, which can be caused by sugarcane mosaic virus (SCMV), Sorghum mosaic virus (SrMV), Maize dwarf mosaic virus (MDMV), Johnsongrass mosaic virus (JGMV), Zea mosaic virus (ZeMV), and *Sugarcane streak mosaic virus*, (SCSMV) six kinds of the virus of potato virus (SCMV) were isolated or co-infected. The pathogenicity of different virus strains to the same sugarcane variety is different. Therefore, it is necessary to investigate the virus types of sugarcane mosaic disease in the field to determine the ability of new varieties to resist the dominant virus. Pathogen identification techniques include direct detection, electron microscopy, serological detection, molecular identification, and host identification. Chen et al. (2001) identified

SrMV and SCMV as the primary pathogen of sugarcane mosaic disease in Zhejiang Province by molecular biological methods; Li et al. (2001) showed that the pathogen of sugarcane mosaic disease in Fujian was SrMV; the pathogen of sugarcane mosaic disease in Guangdong, Guangxi, and Hainan was SCMV; the most recent research report in Yunnan was SrMV (Wang et al., 2009).

4.3.2.2 Matching of parent combinations

The resistance of sugarcane to mosaic disease was different among different varieties. The test showed that the resistance of hybrid progenies to mosaic disease was related to parents' resistance. The parents had high resistance, and their offspring had high resistance, which allowed them to select resistant varieties to control the mosaic disease. In selecting parent combinations, selecting the dominant virus strains with resistance to mosaic disease and parents with certain yield traits is necessary. Generally, the best parents selected as cross combinations are resistant to mosaic disease, or at least one parent is resistant. It is appropriate to use artificial vaccination or natural infection to identify the resistance of lines with better agronomic traits such as yield and sucrose score and screen out the resistant varieties in the breeding process.

4.3.2.3 Identification method of sugarcane resistance to mosaic disease

Resistance identification can be divided into natural inoculation method and artificial inoculation method.

(1) Artificial inoculation

Planting of test clones: the sugarcane seeds of the tested sugarcane clones were cut into single bud seedlings and then planted in the field (or basin) after accelerating germination. The direct seeding method was adopted, and 30 seedlings were ensured for each material.

A. Inoculation method

The virus inoculation solution was prepared by adding 3 times phosphoric acid buffer solution (containing 0.2% Na_2SO_4) with a concentration of 0.1 mol/L and pH of 7.2 into the infected leaf tissue, grinding for 3 min, filtering, and squeezing with double-layer gauze. Inoculation operation: 500–600 mesh quartz sand was placed at the base of the heart leaf of sugarcane seedlings at the 2–3 leaf stage. The thumb and index fingertips were moistened with inoculation solution, and the base of the heart leaf was pressed and rubbed 2 or 3 times to scratch the leaf epidermis.

B. Investigation record items

The date of vaccination, the number of inoculated seedlings, the onset of disease symptoms, and the cumulative number of clusters were investigated every 15 days until the disease was stable.

(2) Natural infection

Sugarcane areas where the mosaic disease is easy to occur and spread were selected as natural susceptible test fields. The tested clones were randomly arranged in each plot and repeated. A row of infected sugarcane plants was planted every two lines of materials as the pathogen, and representative resistant varieties and susceptible varieties were planted as control. The disease resistance of the identified materials was determined by comparing with the control.

4.3.2.4 Classification standard

The incidence rate has 5 grades: the first level is immunity, the incidence rate is 0; the two-level is high resistance, the incidence rate is 1% to 10%; the three-level is intermediate resistance, the incidence rate is 10.1 % to 33 %; four grade is the sensitive disease, the incidence rate is 33.1 % to 66 %; the five grade is the high sense, the incidence rate is more than 66.1 %.

4.3.2.5 Achievements and progress of breeding for resistance to mosaic disease

The major sugarcane breeding organizations in China pay more attention to the breeding and identifying sugarcane varieties and germplasm resources with resistance to mosaic disease as one of the evaluation indexes. Many sugarcane varieties and germplasm resources with resistance to mosaic disease have been bred and identified. Li et al. (2009a; 2009b) used the highly pathogenic Sorghum mosaic virus Isolate (SrMV-HH1) to identify and evaluate sugarcane varieties, lines, and resources, 'CYR99-155' 'CYZ89-351' 'CYZ99-596' 'CGZ95-108' 'RB72-

454' and other varieties showed level 1 immunity; 'Q170' showed level 2 high resistance; 'SP71-6180' 'CDZ93-34' 'CYZ02-588' and 'CYZ99-91' showed grade 3 intermediate resistance. 'CYZ98-13' 'CYZ00-45' 'CYZ01-1029' 'CYZ99-113' 'CYZ02-1826' and 'CYZ98-236', showed grade 1 immunity among the excellent clones. 'CYZ01-347' and 'CP1' showed grade 3 intermediate resistance. Among the excellent intermediate clones, 'CYZ03-113' 'CYZ03-313' and 'CYZ03-314' showed grade 1 immunity, while 'CYZ03-311' and 'CYZ03-316' showed grade 3 intermediate resistance. Among sugarcane resources, '51NG92' '96NG16' '57NG155' 'Muckche' '48Mouna' '27MQ1124' 'Luohan sugarcane' 'Yundian95-20' ('CYD95-20') 'CHN92-84' and 'F172' showed grade 1 immunity, while Cattle Sugarcane showed grade 3 intermediate resistance. Because there are many kinds and races of the mosaic virus, it is impossible to use the local dominant virus for inoculation tests in all sugarcane areas. Therefore, although some excellent sugarcane varieties are susceptible to mosaic disease, they may be resistant to other mosaic virus species or races in other sugarcane areas, and virus-free healthy seedlings can be used for trial planting in other sugarcane areas.

Many researchers introduced the coat protein (CP) gene of the mosaic virus into sugarcane and obtained resistant transgenic sugarcane seedlings in sugarcane mosaic virus resistance transgenic research. For example, Yao et al. introduced the coat protein gene (ScMV-CP) of sugarcane mosaic virus (ScMV-CP) into 'Badila' sugarcane species susceptible to mosaic disease. They obtained resistant transgenic sugarcane seedlings (Gan et al., 2013).

4.3.3 Breeding of sugarcane varieties to rust resistance

Sugarcane rust was first discovered in Java in 1890. The disease has occurred in India, Cuba, Jamaica, Mexico, and other countries, resulting in a severe reduction in sugarcane production. In 1978, the disease was found in northern Queensland, Australia, and soon spread throughout the Australian sugarcane area. The yield loss was 10%–20%. In recent years, the yield loss has become more and more serious.

In 1978, severe brown rust occurred in Florida of the United States, which made the high resistant variety 'CP70-1133' lose resistance; in 1987, the moderately susceptible variety 'CP72-1210' suffered from the disease, with a yield loss of 20%–25%, and an economic loss of about 40 million US dollars; in 1988, 'CP78-1247' was seriously damaged, with a yield loss of 40%. The disease has now spread throughout the United States. More than 20% of the planting area in central Thailand was severely affected by brown rust in 1991. In 1995, it expanded to Mauritius in Africa.

Sugarcane rust first occurred in China in 1977, when F176 in Taiwan Province was seriously infected by rust. It occurred sporadically in Changning and Gengma of Yunnan Province in 1982. Sugarcane rust was reported in Guangdong Province in 1985. It was identified as Puccinia nigra. In 2009, the sugarcane rust of Taitang86-1626 planted on Binhai farm in Beihai, Guangxi, was severe. The incidence rate of disease leaves was 80 %, and the rate of disease stalks was 100% in serious fields. Sugarcane rust has been widespread in Yunnan, Fujian, Sichuan, Jiangxi, Guangdong, and Guangxi. Sugarcane yield was generally reduced by 15 % to 30%, the weight of sugarcane was more than 40%, and the sucrose content was reduced by 10%–36% (Liu et al., 2007).

4.3.3.1 Pathogeny

Puccinia kuehnii Butler, *Puccinia erianthi* Padw.et Khan and *Puccinia melanocephala* H.Sydow were reported as pathogens of sugarcane leaf rust at home and abroad, *Puccinia kuehnii* was found in Africa, Asia (China, India, Thailand) and Australia, the United States, the Caribbean, and the Atlantic Islands; *Puccinia erianthi* was found just in China and India. However, it is considered that *Puccinia erianthi* and *Puccinia melanocephala* are the same pathogens in foreign countries. At present, more works of literature use *Pucciniamelanocephala*. *Puccinia kuehnii* is the pathogen of yellow rust. It is a kind of sudden pathogen, which cannot expand to an epidemic scale. *Puccinia erianthi* and *Puccinia melanocephala* are the pathogens of brown rust, which are harmful to sugarcane. They often cause the occurrence and epidemic of sugarcane rust and have led to the elimination of cultivated varieties 'B4362' and 'Co475'. Sugarcane rust is an air-borne disease. Breeding and popularizing resistant varieties are the most economical and effective measures to control the disease.

4.3.3.2 Matching of parent combinations

Deng et al. (1994) and Wang et al. (1994) evaluated the rust resistance ability of commercial sugarcane varieties and their hybrid progenies by studying the response of the populations to rust. The results showed that the heritability of rust resistance reached 0.7–0.8, and the hybrids showed high resistance frequency; the results showed that the rust grade of the generation population tended to the distribution of disease resistance, which indicated that it was feasible to select clones with resistance to rust through sexual hybridization. Therefore, it is necessary to identify the disease resistance of sugarcane germplasm resources and select the cross combinations of resistant parents. The parents of cross combinations should choose the rust-resistant parents, or at least one resistant.

4.3.3.3 Breeding method of sugarcane rust resistance

In the breeding process of rust-resistant varieties, the screening of rust-resistant clones should be strictly carried out. The incidence of sugarcane hybrids should be observed at all stages of the breeding process. Significantly, the rust resistance identification of sugarcane varieties entering the regional test should be carried out in different places and years. The evaluation conclusion of the disease resistance of sugarcane varieties is more reliable.

To delay the loss of disease resistance, we should pay attention to the rational distribution of disease-resistant varieties and avoid large-scale cultivation of a single variety.

4.3.3.4 Identification method of sugarcane rust resistance

(1) Natural infection

If there are plenty of pathogens around the test area, the natural infection method can be used, and the species of rust pathogens can be identified. The field experiment was randomized block design, 3 replicates, seedling transplantation, 2 rows per plot. Among them, 30 plants were planted in two rows for each tested clone; simultaneously, the highly susceptible clones were used as the inducer of disease, and 15 plants were planted in each plot, which was also used as the susceptible control. Trial management is the same as field management and keeps the sugarcane field moist.

Data collection method: three surveys were conducted in the early, middle, and late stages of rust occurrence. 5 stalks in each row and 10 stalks in each plot were investigated. Investigation items: ① The average number of spores in the same leaf was calculated from the number of spores in the second 1/3 area of a leaf; ② The percentage of the infected area of leaves in the highest visible hypertrophic zone was estimated by visual observation; ③ Visually estimate the severity of rust on the whole plant. The classification standard of sugarcane rust is shown in Table 4-1.

Table 4-1 Classification standard of sugarcane rust

Description	Reaction of infection	Resistance response
0	No symptoms	HR
1	There were a few isolated spots on the tip of a few older leaves	R
2	The disease spots were increased on the older leaves	R
3	In most of the old leaves, there were many agglomerated lesions, and there were small sores in the center of necrotic lesions	MR
4	In almost all the old leaves, large agglomerated spots and sores were added	MR
5	In many old leaves and some new leaves, there are many spots and larger sores	S
6	In the new young leaves, there are large agglomerated spots and sores	S
7	Most of the new leaves are covered with a large number of agglomerated lesions and sores	S
8	Lot of sores and agglomerations on lobus cardiacus	HS
9	The plant is seriously infected as if it had been burned by fire	HS

Note: HR: Highly Resistant; R: Resistant; MR: Moderately Resistant; S: Susceptible; HS: Highly Susceptible.

(2) Artificial inoculation

If there are few pathogens around the test area, artificial inoculation can be used. Five barrels per clone, 5–8 plants per barrel, 30 plants per clone for inoculation, routine management.

The dominant sugarcane rust fungus spores in the local sugarcane area were used as the inoculation pathogen. The inoculum was inoculated with 40–50 field spores and sprayed. After injection, the sugarcane plants were cultured in the shading net shed and watered 2 or 3 times a day to keep the sugarcane moist. After 28–35 days of the disease, the disease degree was investigated.

The investigation record items are the same as those of natural infection.

4.3.3.5 Classification standard

The classification standard of sugarcane rust is shown in Table 4-1. The occurrence degree of rust was different among different clones. In the early stage of the epidemic, the difference between clones may not be noticeable. In the middle and late stages of the epidemic, the more serious the rust, the more pronounced the difference. Therefore, it is essential to create favorable conditions for the occurrence and development of rust resistance tests. According to the occurrence characteristics of rust, besides the resistance of the host, the most critical environmental factors are temperature, humidity, and water.

4.3.3.6 Achievements and progress of rust resistance breeding

The main sugarcane planting countries attach great importance to the breeding of sugarcane rust resistance. In the breeding process, there are identification and screening procedures for rust resistance breeding. At the same time, they attach great importance to screening resistant sources and reasonably selecting resistant parent combinations. Some resistant materials have been screened out in China. Li et al. (2008) screened and evaluated sugarcane germplasm resources with the spores of *Puccinia erianthi* as an inoculation source. The evaluation results showed that: among the *S.officinarum* species '48mouna' 'Kara karawa' 'Muckche' 'Badila', and '96NG16' were grade 1, showing high resistance; '51NG103' 'NC32' 'Cana Blanca' 'Guam A' 'Keong Java' were grade 2 resistant, '57NG155' 'Nagans' 'Crystalina' were grade 3 moderately resistant. The '*S.sinense*' 'Jiangxibamboo cane' 'Guangxibamboo cane' 'Sichuan Lucane' 'Wenshan cane' 'Heqing strawing cane' 'Pansahi' 'Uba' were grade 1 high resistance; Guangdong bamboo cane was grade 2 disease resistance; Yongsheng cane and Henan Xuchang cane were medium resistant. Among the *S.barberi*, 'Hatuni' was grade 1 high resistance, and 'Nagori' was grade 3 medium resistance.

In recent years, with the rapid development of molecular biology techniques, the application of molecular biology techniques in sugarcane breeding for rust resistance has made some remarkable progress. Grivet et al. (Que et al., 2005) found a rust resistance gene linked to the probe CDSR29; Asnaghi et al. (2000) found a marker linked to the rust resistance gene on the map; Barne et al. (1998) found two specific RAPD bands that can be repeated between the resistant and susceptible genotypes through RAPD analysis of 8 rust-resistant lines and 8 rust susceptible mutants of sugarcane variety 'Nco376'. One of them has been cloned, sequenced and transformed into SCAR marker. Molecular marker-assisted selection of rust resistance genes can significantly speed up the conventional breeding process and improve breeding efficiency.

References

ASNAGHI C, PAULET F, KAYE C, et al., 2000. Application of synteny across poaceae to determine the map location of a sugarcane rust resistance gene. Theoretical and Applied Genetics, 101 (5-6):962.

CHAO C P, HOY J W, SAXTON A M, et al., 1990. Heritability of sugarcane smut resistance and correlation between smut grade and yield components. Crop Science (23):54-56.

CHEN R K, LIN Y Q, ZHANG M Q, et al., 2003. Theory and practice of modern sugarcane breeding. Beijing:China Agricultural Press (in Chinese).

DENG Z H, LIN Y Q, WANG J N, et al., 1994. The Inheritance of resistance to rust and breeding strategy in sugarcane II. The analysis of combining ability for rust resistance of parental combination. Journal of Fujian Agricultural University, 23 (3):249-252.

GAN Y M, ZHANG S Z, ZENG F Y, et al., 2013. Advances in sugarcane transgenic breeding. Biotechnology Bulletin (3):1-9.

GLORIA B A, CAPOTE ALBERNAS M C, AMORIM L, et al., 1999. Morphological characteristics of sugarcane clones, susceptible and resistant to smut (Ustilago scitaminea). In:Rao GP, Bergamin FA, Magarey RC, et al. 'Sugarcane pathology Vol 1:Fungal diseases'. Enfield:Science Publishers, Inc:167-182.

GONG D M, LIN Y Q, CHEN R K, 1996. Studies on breeding techniques for smut Resistance in sugarcane III. Relationship between bud characteristics and resistance to smut. Acta Agronomica Sinica, 22 (3):362-364.

GU L H, ZHANG S Z, YANG B P, et al., 2008 Introduction of chitin and β-1, 3-glucan into sugarcane. Molecular plant breeding, 6 (2):71-74.

HUANG J Y, HE H, BI S L, et al., 2001. Screening of superior sugarcane lines resistant to smut. Guangxi Sugar Industry (1):6-8.

INGELBRECHT I L, IRVINE J E, MIRKOV T E, 1999. Posttranscriptional gene sile-ncing in transgenic sugarcane. Dissection of homology-dependent virus resistance in a monocot that has a complex polyploidy genome. Plant Physiology, 199 (4):1187-1198.

KONG Y, 2012. The study of KP4 gene genetic transformation of sugarcane. Haikou:Hainan University.

LI Q W, CHEN Z Y, LIANG H, 1999. Modern sugarcane improvement technology. Guangzhou:South China University of Technology Press (in Chinese).

LI W F, CAI Q, HUANG Y K, et al., 2008. Identification of sugarcane cultivated original species resistance to *Puccinia erianthi*. Journal of Yunnan Agricultural University, 23 (1):25-28.

LI W F, HUANG Y K, LU W J, et al., 2009a. Resistance evaluation of sugarcane germplasm resources to SrMV. Journal of Yunnan Agricultural University, 24 (3):361-363,373.

LI W F, HUANG Y K, LUO Z M., et al., 2009b. Identification and evaluation of fine sugarcane varieties and samples resistant to SrMV-HH1. Journal of Southwest Agricultural, 22 (1):92-94.

LI Y R, 2010. Modern sugarcane science. Beijing:China Agricultural Press (in Chinese).

LIU X M, LIU W B, SHI H H, et al., 2007. Research progress on sugarcane rust. Chinese plant protection, 27 (12):12-15.

MCQUALTER R B, DALE J L, HARDING R M, et al., 2004. Production and evaluation of transgenic sugarcane containing a Fiji disease virus (FDV) genome segment S9-derived synthetic resistance gene. Crop and Pasture Science, 55 (2):139-145.

ORLANDO B H, THOMMA B P, CARMONA E, et al., 2005. Identification of sugarcane genes induced in disease-resistant some clones upon inoculation with *Ustilago scitaminea* or *Bipolaris sacchari*. Plant Physiol Biochem, 43:1115-1121.

PADMANABAN P, ALEXANDER K C, SHANMUGAM N, 1988. Studies on certain characters associated with smut resistance. Indian Phytopathology, 41:594-598.

QUE Y X, JIANG K Q, XU L P, 2005. Marker-assisted selection in plant breeding for disease resistance and sugarcane. Letters in biotechnology. (4):452-455.

QUE Y X, LIN J W, ZHANG J S, et al., 2008. Molecular cloning and characte rization of a non-TIR-NBS-LRR type disease resistance gene analogue from sugarcane. Sugar Tech, 10:71-73.

QUE Y X, XU L P, LIN J W, et al., 2009. Application of *E.arundinaceus* cDNA microarray in the study of differentially expressed genes induced by *U.scitaminea*. Acta Agronomica Sinica, 35 (5):940-945.

WALKER D I T, 1980. Heritability of smut resistance. Sugarcane Breeders'.

Newsletter, 43:19-23.

WANG J N, LIN Y Q, DENG Z H, et al., 1994. The inheritance of resistance to rust and breeding strategy in sugarcane I. Analysis of resistance to rust in F_1 population. Journal of Fujian Agricultural University, 23 (2):140-144.

WANG X Y, LI W F, HUANG Y K, et al., 2009. Research progress on sugarcane mosaic disease. Sugar Crops of China (4):61-64.

WU K K, LADD S L, HEINZ D J, et al., 1977. Combining ability analysis in sugarcane smut resistance. Sugarcane Breeder'. Newsletter, 39:59-62.

XIA H M, HUANG Y K, WU C W, et al., 2009. Application research on evaluation system for smut resistance identification from Australia for disease resistance breeding in Yunnan. Southwest China Journal of Agricultural Sciences, 22 (6):1610-1615.

XIA H M, WU C W, CHEN X K, et al., 2007. Evaluation of some sugarcane genotypes for smut resistance. Sugar Crops of China (3):26-27.

XU L P, CHEN R K, 2000. Current status and prospect of smut and smut resistance breeding in sugarcane. Fujian Journal of Agricultural Sciences, 15 (2):26-31.

XU L P, CHEN R K, PAN D R, et al., 2001. Analysis of sugarcane segregating population and construction of pools resistance or sensitive to smut. Journal of Fujian Agricultural University, 30 (2):153-157.

ZHANG X, 1998. Crop ecological breeding. Beijing:China Agricultural Press (in Chinese).

5 Stress Resistance Breeding and Its Effect in Sugarcane

5.1 Impact on Cane Sugar Industry of Main Natural Disasters in China

Sugarcane is native to the tropics and subtropics. It is a crop that likes high temperature, humidity, and strong light. Sugarcane areas in China are widely distributed, with complex natural and geographical conditions, a strong three-dimensional climate in some areas, large mountainous areas, and distributed in plateaus, hills, basins, and river valleys. The area has high altitudes, large drops, and diverse landforms. In addition, climate warming and frequent natural disasters have seriously impacted the development of sugarcane production. Drought and freezing injury have the greatest impact on sugarcane production in China's main sugarcane producing areas.

5.1.1 Effects of drought on the cane sugar industry

Since the 1970s, the regional distribution of sugarcane in China has shifted from the southeast coastal areas with suitable light, temperature, and water conditions to the arid, barren and harsh environmental conditions in southwest China. At the same time, with the development of China's economy, affected by the impact of economic crops such as fruits, vegetables, flowers, tobacco, etc., the area of traditional sugarcane suitable areas has been continuously reduced, and sugarcane production has gradually developed to dry slope land without water shortage or irrigation conditions. In addition to the continuous destruction of the ecological environment, drought is becoming increasingly severe. Spring drought, summer drought, autumn drought, and winter drought often occur. The losses caused by drought in agriculture are more than the sum of other natural disasters, seriously hindering the development of agriculture. The area of sugarcane planting in dryland accounts for more than 85 % of sugarcane planting in China. The drought in winter, spring, and autumn has adverse effects on the sowing, sprouting, tillering, and elongation of sugarcane, leading to the poor growth of sugarcane and reducing yield and sugar content, which restricts the development of the sugarcane industry. Therefore, cultivating new sugarcane varieties with high yield, high quality, and drought resistance and improving the water use efficiency of sugarcane are the core technologies to improve the international competitiveness of China's sugarcane industry.

Since the autumn of 2009, the two largest sugarcane areas in Guangxi and Yunnan in southern China have suffered the worst drought in the century. Due to the lack of drought-resistant varieties, the output of sugarcane raw has been reduced by 10.1021 million tons, and the output of cane sugar has been reduced by 2.258 million tons. Among them, the production of sugarcane raw in the Yunnan sugarcane area has been reduced by 4 million tons; The output of cane sugar has been reduced by 460 thousand tons, resulting in more than 4 billion yuan loss in Yunnan's whole cane sugar industry. Therefore, taking strong measures to increase the cultivation of drought-resistant sugarcane varieties and large-scale popularization and application on dry slope land can not only promote the recovery and development of the Yunnan cane sugar industry as soon as possible but also increase the ability of the cane sugar industry to resist natural drought and ensure the healthy and stable development of Yunnan cane sugar industry.

5.1.2 Effects of freezing injury on the cane sugar industry

With improved sugarcane varieties, sugarcane planting has been expanded from tropical areas to subtropical and temperate regions. Still, sugarcane is often affected by low temperature and freezing injury in subtropical and temperate regions. Low temperature seriously affects the yield and quality of sugarcane (Yang, 1999), especially in recent years, climate change is abnormal, extremely low temperature frequently occurs, sugarcane is affected by cold disasters, which causes significant economic losses to sugarcane farmers and sugar enterprises (Li, 2010). New industries have been developing continuously in the advantageous sugarcane-producing areas in southwest Yunnan Province of China. Oriental tobacco, potato, and other industries have shown substantial competitive advantages. Traditional crops such as rice, vegetables, and fruits also have certain comparable benefits compared with planting sugarcane. Sugarcane growing areas are constantly shifting to mountainous areas. The planting altitude of sugarcane has risen from 1,300 m to 1,600 m, and some sugarcane areas have even reached 1,800 m. In addition, due to the lack of rain and drought in winter and spring, the surface soil and plants at night are strongly cooled, and low temperature and frost hazards occur from time to time. Most of them occur in late December and early January of the following year. The low-temperature frost in the spring sowing period can easily lead to the freezing death of seed buds, low seedling emergence rate, and insufficient essential seedlings, resulting in the decrease of sugarcane yield and total yield, the poor harvest of sugarcane farmers, and the shortage of raw materials for sugar production in sugar factories; the low temperature and frost damage at the mature stage of sugarcane will lead to the decrease of sugar content, the reduction of yield and income of sugar mills, as well as the difficulty of seed retention, the low seeding rate of ratoon crop, and the deterioration of seedling quality. Therefore, it is of great significance to strengthen the research and utilization of sugarcane varieties and cold-resistant germplasm resources, breeding, and reasonably distributing cold-resistant varieties to ensure the stable development of sugarcane production and the cane sugar industry.

5.2 Drought Resistance Breeding

Sugarcane is a tropical and subtropical crop with the characteristics of high temperature, significant water demand, fertilizer absorption, and extended growth period. The whole growth and development process needs higher temperatures and abundant rainfall. Generally, the active accumulated temperature above 10 °C is 5,500–6,500 °C, the annual sunshine hours are more than 1,400 hours, and the annual rainfall is more than 1,200 mm. Water is one of the most critical factors in sugarcane production. According to statistics, about 30% of sugarcane production depends on the appropriate water supply (Zhang, 2011).

5.2.1 Water requirement of sugarcane

5.2.1.1 Water absorption form of sugarcane

The water absorption mechanism can be divided into two forms: passive water absorption and active water absorption.

Passive water absorption: the main form of water absorption by sugarcane roots. The above-ground part of the plant is deficient in water due to transpiration. To meet the water supply of the above-ground part, the plant transports the root water to the stems and leaves above the ground through the dredging tissue, resulting in the water shortage of root cells and the decrease of water potential, thus causing the root cells to absorb water from the soil.

Active water absorption: it has nothing to do with the above-ground plant activities but is caused by the metabolic activities of sugarcane roots. Water absorption power is the root pressure produced by the physiological activities of the root system. The size of root pressure depends on the concentration difference between root vessel sap and soil solution and the water permeability of root cells.

5.2.1.2 Water requirement of sugarcane growth

The water consumption of sugarcane is composed of ecological water consumption and physiological water demand. Ecological water consumption refers to the water loss caused by soil evaporation in the field. Long growth periods, high temperature, less rainfall, low relative air humidity, and insufficient soil water holding capacity lead to significant water consumption. Physiological water requirement refers to the water absorbed by sugarcane roots and used for plant construction and water loss due to transpiration. The water requirement of sugarcane at different growth stages is very different. The general trend is "less in the early stage, more in the middle stage, and less in the later stage", that is, the water requirement in the germination stage is the least, the tillering stage is gradually increasing, the water demand in the elongation stage is the largest, the water demand at the end of elongation is gradually reduced, and the minimum is at the mature stage.

(1) Water demand in germination period

The germination ability of sugarcane is closely related to the water content of sugarcane buds. When the water content of sugarcane seeds is less than 50%, the germination ability is weakened; when the moisture content of sugarcane seeds is about 70%, it can be used for germination (Li, 2010; Zhang, 2011). However, to promote the germination and development of sugarcane seeds, soaking seeds are usually used to supplement the water content of sugarcane seeds. The water requirement of sugarcane in the germination stage is not much. The water supply in this period is mainly ecological water for seedling germination, accounting for 8.4%−18.1% of the total water demand in the whole growth period. At this stage, the water content in the 25 cm range of the soil surface must be 55%−70% of the maximum water holding capacity, and less than 55%, which often leads to fewer sugarcane seedlings due to drought.

(2) Water requirement at tillering stage

The tillering period of sugarcane is about 2 months. Due to the increase of sugarcane leaves, the water demand also gradually increases, accounting for 15.4%−21.7% of the total water demand in the whole growth period. Ecological water demand is still the primary water demand. It is generally required that the moisture content of the soil surface within 30 cm of the maximum water holding capacity is 65%−80%, in which the soil moisture content is 70%. The content lower than 65% or higher than 80% is not conducive to the growth of sugarcane, if the field ponding, poor soil ventilation, it will make the leaves discolor, growth hindered, and even rot death.

(3) Water demand in elongation period

The elongation stage is the most vigorous period of sugarcane growth. After the sugarcane enters the elongation period, the sugarcane has enormous growth, large leaf area, deep root system, and water absorption capacity. Transpiration and photosynthesis are the strongest, and the physiological activities such as synthesis, transportation, and accumulation of organic matter need a lot of water. At this stage, physiological water consumption was the primary water consumption, accounting for more than half of the total water demand in the whole growth period, about 54.3%–57.8%. The water content within 50 cm of the soil surface layer should reach 80%–90% of the maximum water holding capacity.

(4) Water demand in mature period

In the mature stage of sugarcane, the growth rate of sugarcane was slowed down, the sucrose accumulation was accelerated, and the water requirement of sugarcane was reduced. This stage is suitable to keep the water content within 40 cm of the surface soil accounting for 60%–70% of the maximum field water holding capacity.

5.2.2 Drought characteristics of sugarcane growing areas in China and the demand for drought resistance of sugarcane varieties

Firstly, because of the unpredictability of drought occurrence and duration, drought varieties in different years is different. Therefore, the varieties suitable for planting in dryland sugarcane areas must have good yield potential in normal years and also be able to endure a certain degree of drought environment. When affected by drought, the yield loss is slight. That is, it has a good high and stable yield performance under various water conditions. It does not mean that the above varieties can be planted under extreme drought conditions, and any varieties will be inhibited or even died under severe drought conditions. Secondly, there are differences in the occurrence of drought in different sugarcane areas, and the requirements for drought resistance of sugarcane varieties are also different. The practice has proved that to obtain a high yield of sugarcane in the dry land, breeding drought-resistant and high yield varieties is the most economical and effective means besides scientific drought-resistant cultivation methods. When evaluating the drought resistance of sugarcane varieties in production, the sensitivity of sugarcane varieties under drought conditions and the growth compensation ability after drought relief should be fully considered. Finally, the degree of affected growth performance and yield should be taken as the evaluation standard of drought-resistant varieties.

5.2.3 Drought resistant breeding of sugarcane

5.2.3.1 Drought resistance of sugarcane

The loss of crops caused by drought is the first abiotic stress, second only to biological stress caused by diseases and insect pests. Many sugarcane varieties have various drought adaptation modes, including drought avoidance, drought tolerance, and drought resistance. Drought tolerance and drought resistance are collectively referred to as drought resistance (Figure 5-1).

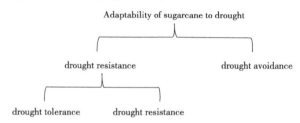

Figure 5-1 Adaptation of sugarcane to drought

(1) Drought resistance of sugarcane

Drought resistance belongs to morphological drought resistance, which means that underwater stress, sugarcane can maintain a high water potential by limiting water loss and maintaining some water absorption capacity to keep various physiological and biochemical processes in a normal state. As for the drought resistance of sugarcane, since the 1930s, many researchers have studied leaf morphology, root development, and water transport capacity. Evans (1935) pointed out that the drought-resistant sugarcane varieties were characterized by few leaves, narrow and thick erect, high degree of keratinization of epidermal cells, depression of stomata, and more and longer setae. Subsequent studies showed that thickening of the cuticle, widening of vesicular cell band, more veins, and fewer stomata per unit area were related to drought resistance (Xu, 1986). The developed degree of the root system, physiological activity, and active water absorption capacity is also closely related to drought resistance (Evans, 1937; Weller, 1931). Therefore, the varieties with strong drought resistance have more significant root pressure and bleeding volume. In addition, the number of vessels per unit area in the root and stem of sugarcane was positively correlated with drought resistance. The development of mechanical tissue around the vascular bundle of drought-resistant varieties was poor, and the degree of lignification of thick-walled cells was higher (Tan, 1988).

(2) Drought tolerance of sugarcane

Drought tolerance of sugarcane is mainly realized by plasma membrane permeability, osmoregulation, and the ability of antioxidant enzymes to scavenge free radicals (Chen et al., 1994; Liang et al., 1995; Zhong et al., 2002). In other words, sugarcane can maintain cell swelling under low water potential to provide physical strength for plant growth under severe water stress and maintain protoplast and its principal organs in severe dehydration with minor damage and remain unaffected, which belongs to the category of physiological and biochemical drought resistance.

Under drought stress, the permeability of the plasma membrane of sugarcane leaves increased continuously. The increase of drought-resistant varieties was lower than that of non-drought-resistant varieties, and the damage of plasma membrane was less. The damage of the membrane system caused by membrane lipid peroxidation under drought stress is mainly determined by measuring the content of malondialdehyde (MDA) produced by the peroxidation of unsaturated fatty acids in membrane lipid the content of MDA is positively correlated with the change of plasma membrane permeability. Proline accumulation is considered an effective index of drought resistance in sugarcane. Under drought stress, the substances involved in osmoregulation can be divided into two categories: the organic solutes synthesized in cells and the accumulated compatible solutes mainly include mannitol, proline, glycine, betaine, trehalose, fructan, inositol, polyamine, and other small molecular compounds. Under drought stress, the osmotic potential of sugarcane leaf cells decreased. At the same time, the concentration of reducing sugar and potassium increased, which indicated that sugarcane could be osmotically regulated by increasing the concentration of cell solute under drought stress.

5.2.3.2 Identification of drought resistance in sugarcane

Under drought conditions, soil water content decreased, and the water supply of sugarcane was limited. Drought stress first causes changes in the structure and composition of cell biofilm. It then damages cellular osmoregulation and antioxidant pathways, resulting in excessive free radicals and membrane lipid peroxidation, resulting in cell injury. Under drought stress, the relative water content of sugarcane leaves decreased significantly. The natural saturation deficit intensified, the chlorophyll content of sugarcane leaves decreased, the MDA content and the plasma membrane's relative permeability increased. The results showed that the photosynthetic capacity of sugarcane was damaged by water deficiency, which affected the electron transport and CO_2 assimilation process of photosynthesis. The photoreduction activity of chloroplast decreased. The chlorophyll fluorescence parameters FV-FM and FV-FO were significantly decreased. Drought stress reduced the primary light energy conversion efficiency, net photosynthetic rate, transpiration rate, proline content, and catalase activity (Luo et al., 2000; Ye et al., 2003).

With the increase of water stress, the damage degree of sugarcane was deepened, and the number of leaves, plant height, fresh weight, and dry weight of the plant decreased. Luo et al. (2005) determined the Leaf proline content, leaf cell membrane permeability, and leaf malondialdehyde content of three sugarcane varieties with different drought resistance under drought conditions. The results showed that the proline content and plasma membrane permeability were significantly correlated with the drought resistance of sugarcane varieties. These physiological indexes are closely related to the drought resistance of sugarcane, reflecting the drought resistance of different sugarcane genotypes. They can be used as identification indexes in sugarcane drought resistance breeding and directly guide the drought resistance breeding of sugarcane. In recent years, there has been a new exploration in the identification of sugarcane drought resistance. Zeng et al. (2003) tested 30 sugarcane germplasms with physiological indexes such as the water retention rate of detached leaves, the relative water content, and stomatal opening. The results showed that these indexes were consistent with the actual drought resistance performance in production.

Drought resistance is a complex quantitative character affected by many factors. Different varieties have different responses to a single index. Therefore, it is difficult to use a single index to comprehensively reflect the drought resistance of various varieties, so it is necessary to use multiple indexes combined with a statistical method to conduct a comprehensive evaluation. Wu et al. (2011), Gao et al. (2002; 2006), Chen et al. (2007) analyzed and evaluated the drought resistance of sugarcane varieties and hybrid combinations by combining drought resistance physiological index, yield character index, and statistical means, which increased the authenticity and reliability of sugarcane drought resistance evaluation.

5.2.3.3 Breeding and selection of drought-resistant sugarcane varieties

(1) Evaluation and utilization of drought-resistant sugarcane germplasm resources

Drought resistance evaluation of germplasm resources is the basis of sugarcane drought resistance breeding. The cultivation of drought-resistant varieties depends on the exploration and utilization of drought-resistant parents. Because of the narrow genetic basis and poor drought resistance of sugarcane varieties in drought resistance breeding, sugarcane breeders have carried out extensive screening and evaluation of sugarcane drought-resistant germplasm resources from the perspective of systematic development and achieved remarkable results. Pan et al. (2006) evaluated the drought resistance of five sugarcane species and their derivatives by using the established drought resistance physiological identification technology and screened several drought-resistant germplasm materials for sugarcane breeding, such as Jiangxibamboo cane, 'NCo310' 'ROC5', Fujianbamboo cane, 'CGT11' 'CYZ89-351' 'CGT89-5' 'CFN81-745' 'CYC90-33' and 'CYC73-512', which can be used in sugarcane breeding. 'CYR95-128' and 'CYR99-719' were selected by Jing et al. (2002) as parents for drought resistance breeding. Yang et al. (2008) evaluated the related drought resistance of the progenies of the combinations of Yunnan Xishuangbanna *S.spontaneum*, Yunnan Manhao *S.spontaneum,* and Yunnan Funing *Erianthus arundinaceus*. The results showed that Yunnan Funing *Erianthus arundinaceus* had the strongest drought resistance, followed by the progenies of Yunnan Manhao *S.spontaneum* and Yunnan Xishuangbanna *S.spontaneum*, and the germplasm of Yunnan Funing *Erianthus arundinaceus* showed substantial genetic superiority in drought resistance.

Ruili National Inland Sugarcane Hybrid Breeding Station (RSBS) carried out many studies on heritability and combined parents' ability according to the seedling performance of hybrid progenies and selected many excellent parents with high heritability and strong combining ability. The drought resistance of ROC and CYR series parents was evaluated, 'CYR99-131' 'CYR99-601' 'CYR05-283' 'ROC11' 'ROC23' 'ROC24' 'ROC25', and other excellent drought-resistant parents will be bred more and better drought-resistant sugarcane varieties through continuous cross-utilization. It was also found that the drought resistance of some parents, such as 'CP72-1210' and 'ROC1' derived varieties, depended on the drought resistance of another parent: 'YN73-204' 'CYT93-159' and 'CFN95-1702', which were derived from 'CZZ74-141' and 'YN73-204', respectively. The drought resistance of 'CFN91-4710' and 'CFN94-0403' selected from non-drought resistant parent 'Ke5' and 'CMT69-243' were medium and low resistance; 'CMT92-505' and other varieties derived from the combination 'Xuan 15' had weak drought resistance, so they should be used less in drought resistance breeding.

At present, in the field of sugarcane drought resistance breeding research and the existing sugarcane variety resources for combination hybridization, but also actively explore new drought-resistant resources. Previous studies have shown that germplasm materials with the higher genetic relationship of *S. spontaneum* or

E.arundinaceus have strong drought resistance. *E.arundinaceus* is a closely related plant of sugarcane, which has excellent drought resistance ability and been paid more and more attention by breeders. Fu et al. (2009) obtained 19 clones selected from the backcross of 'CYC01-87' (the F_2 of *E.arundinaceus*) and 'ROC20', and the comprehensive performance of some clones was better than or reached the production level of domestic elite varieties.

(2) Evaluation and screening of drought-resistant sugarcane varieties

A. According to the morphological characteristics of drought resistance

In general, the root system of sugarcane varieties with early growth, rapid development, and early closure has a deep distribution and can still obtain higher yields under drought conditions. Sugarcane varieties with deep roots can absorb and utilize the water in the lower soil when the surface soil is short of water, thus effectively avoiding the damage of water stress. The varieties with a high T/A ratio showed no drought resistance, while those with a low (T/A) ratio showed drought resistance.

The transpiration loss of sugarcane plants is closely related to leaf characteristics. The leaf morphology of the drought-resistant type was as follows: short and narrow leaves, low stomatal density and depression, vesicular cells arranged in narrow strips, and thick cuticles. Generally, drought-resistant varieties have fewer stomata and smaller stomata, effectively reducing water loss, affecting gas exchange, and restricting plant growth. It is one of the possible reasons that most drought-resistant varieties are not necessarily high-yield.

B. According to the physiological and biochemical index selection

Proline is a kind of cell-compatible solute, which can affect osmoregulation when the accumulation is significant. When the water content of leaves decreased, all kinds of metabolites would be accumulated in plant tissues, among which the free proline concentration would increase significantly, and the increase of drought-tolerant varieties was more than that of drought-resistant varieties so that proline could be used as one of the indexes for screening drought-tolerant varieties. Under drought stress, MDA content and plasma membrane permeability of drought-resistant varieties were lower than those of non-drought-resistant varieties, which could also be used as screening indexes of drought-resistant varieties. However, the above indicators are not specific to drought stress, and they will change under high temperature and salt stress. Therefore, it is not enough to use only one or two indexes to evaluate the drought resistance of varieties, and multiple indexes must be used comprehensively.

C. Comprehensive evaluation and selection according to growth status

Evaluating and selecting sugarcane lines should be based on production indexes, especially yield, and sugar content. We are still most concerned about whether they can get better harvest under drought conditions for drought-resistant varieties.

Damage rate of plant height: plant height is one of the most critical yield factors of sugarcane. The plant height of sugarcane is more sensitive to water stress, and the damage of water stress to plant height is consistent with drought resistance in production. The injury rate of plant height reflects the degree of plant growth inhibition under drought stress, and it is easy to determine which ideal drought resistance identification index is. The calculation formula is as follows:

$$\text{Damage rate of plant height}(\%) = \frac{\text{Control plant height} - \text{Plant height under drought treatment}}{\text{Control plant height}}$$

Sensitivity to drought stress and response after rewatering: the yield of sugarcane under drought depends on the growth loss during drought and whether the growth can be recovered and recovered after the drought is relieved (rewatered). The response of different sugarcane varieties to drought stress and rehydration was different. Therefore, the response to drought stress and rehydration can be used as reference indexes to evaluate varieties' production performance and drought resistance under drought conditions.

Yield analysis of normal and drought years: due to the unpredictability of drought, it is difficult to objectively evaluate the drought resistance of varieties with the data of one or two years or a few experimental sites during the field test. Therefore, it is necessary to conduct an artificial simulation test and use other indicators for direct or indirect selection. The productive drought-resistant varieties should be those with good yield stability under drought and non-arid conditions. Therefore, the evaluation of drought resistance by directly using the field yield index should be combined with the regional test and productive test of a breeding plan, and the stability analysis of multi-point test data for many years was carried out to evaluate the drought resistance. Attention should be paid to the yield in drought and normal years.

(3) Breeding effect of drought-resistant sugarcane varieties

Sugarcane breeding research institutions of China have conducted many drought resistance evaluations on the existing main and derivative varieties. Many varieties with strong drought resistance have been selected and applied. Under drought stress, 'ROC22' 'ROC20' and 'ROC23' had lower mortality, less loss rate of plant height growth, and drought resistance was significantly better than 'CGT11'; the drought resistance of 'CP80-1827' and 'ROC 21' was also more robust than that of 'CGT11'; the drought resistance of 'ROC16' was moderate and low. At present, many drought-resistant varieties have been screened out, such as 'Guiyin 5' and 'CGT22', which have high cane yield and drought resistance. They are suitable for dry land sugarcane production in Guangxi, and are ideal for both sugar and energy use. Sugarcane Research Institute of Yunnan Academy of Agricultural Sciences (YSRI) is the only provincial sugarcane research institution in Yunnan Province. Because of the continuous transfer of Yunnan sugarcane area to dry slope land and the growing demand for drought-resistant sugarcane varieties in sugarcane areas, the innovation and evaluation of drought-resistant parents are vigorously carried out, and the breeding scale of drought-resistant varieties is significantly increased. Drought-resistant varieties are bred directly on dry land, and many drought-resistant varieties are bred. Among them, 'CYZ 03-194' 'CYZ 05-51' and 'CYZ 08-1609' are outstanding in Yunnan dryland sugarcane area, with an increase of 20−30 t/hm^2

compared with 'ROC22', they have become the main popularized variety in dryland sugarcane area of Yunnan Province. As the participating unit of the National Sugarcane Regional test, the latest sugarcane varieties bred by the national sugarcane research institute were introduced simultaneously. Through the experiment and demonstration, many excellent drought-resistant varieties, such as 'CYT86-368' 'CFN91-21' and 'CGT21' ('CGT94-119'), were successfully selected and planted dryland worst drought in Yunnan in 2009. They show apparent advantages. The further application of these drought-resistant varieties will lay a solid foundation for the sustainable and stable development of the Yunnan cane sugar industry.

5.3 Cold Resistance Breeding

5.3.1 Temperature requirement of sugarcane growth

The suitable temperature for sugarcane germination and growth is about 30 °C. It can germinate above 13 °C, 30–32 °C is the most suitable temperature for germination, and more than 40 °C is unfavorable. Sugarcane roots germinate above 10 °C, and 20–27 °C is the most suitable temperature. The seedlings need to grow above 15 °C and tiller above 20 °C. The optimum temperature for sugarcane stem elongation is about 32 °C, lower than 20 °C, the elongation is slow, and it stops growing below 10 °C. When the temperature exceeds 35 °C, it begins to grow fast and soon slows down; at 40 °C, weak photosynthesis can be carried out. 0 °C is the critical temperature for freezing injury, and the symptoms such as leaf withering and sugarcane stalk withering are easy to appear. The sugarcane growth point will be frozen to death at −2.0–1.5 °C.

5.3.2 Low-temperature damage of sugarcane

5.3.2.1 Classification of sugarcane low-temperature damage
According to low-temperature stress and crop response characteristics, low-temperature damage can be divided into chilling and freezing injuries. Chilling damage refers to the harm caused by low temperatures above 0 °C to crops; freezing injury refers to the harm caused by low temperatures below 0 °C.

(1) Chilling injury

Although sugarcane can survive at a temperature above 0 °C, the growth almost stops when the temperature is lower than 20 °C, and some tissues may be damaged when the temperature is lower than 15 °C. Chilling injury can cause a series of physiological activities of sugarcane plants to be blocked and destroyed, mainly affecting sugarcane germination and sucrose accumulation (Zhang et al., 2000). The mechanism of chilling injury was that the photosynthetic intensity was decreased, the water absorption capacity was weakened, the respiration was abnormal, the permeability of cell membrane was increased, a large number of substances in cells were exuded, the protein decomposition rate in tissues was higher than the synthesis rate, the coordination of many biochemical reactions in tissues was destroyed, and the metabolic activities of sugarcane plants were abnormal (Li, 1998).

(2) Freezing damage

When the temperature dropped below 0 °C, the sugarcane cell tissue was damaged by ice crystals. The mechanical action of ice crystal destroyed the fine structure of cell protoplasm, and the metabolism could not be carried out, which led to the death of the sugarcane plants. Freezing injury can lead to the shooting death of sugarcane shoots, affect the quantity and quality of seedlings, and do more harm to sugarcane production than chilling injury. In the subtropical sugarcane region, frost in autumn and winter can stop sugarcane growth and sugar accumulation and even lead to sucrose transformation in vivo, resulting in sugar loss and reducing the quality of raw sugarcane; frost in spring will affect the planting of new sugarcane, Ratoon germination, and seedling growth, resulting in the shortening of the effective growth period and the decrease of yield.

5.3.2.2 Factors affecting the harm of low temperature

Different sugarcane varieties suffer from different degrees of low-temperature damage due to different cold resistance. The longer the duration of low temperature, the more serious the damage; the lower the temperature, the more serious the damage. The damage caused by freezing damage at 1 °C for 48 h is the same as that caused by −3.8 °C for 1 h. The young tissue and vigorous growing parts had a higher freezing point and were easily damaged, followed by the upper bud and shoot young stem, and the middle and lower bud and stem were the least vulnerable. The results showed that sugarcane had a different freezing injury on different terrain. The mortality rate of sugarcane bud was low in hillside and terrace with high air mobility; it was high in low-lying and valley land with low air mobility; it was seriously frozen in barren and low water content soil; it was light in deep soil, fertile soil, and rich in organic matter and water.

5.3.3 Effects of low temperature on sugarcane growth and development

5.3.3.1 Effect on physiological and biochemical indexes of sugarcane

(1) Enzymatic activity

As regulatory proteins of plant metabolism, Enzymes restrict plants' growth, development, and stress resistance. Under the direct and indirect effects of low temperature, acid invertase activities decreased, while amylase and neutral invertase activities increased. As the critical enzymes of plant metabolism, they are related to the cold resistance of sugarcane, but the correlation is not significant. It needs to be further confirmed that they should be used to index sugarcane cold resistance identification. The increase of peroxidase activity and the number of isozyme bands in sugarcane were beneficial to prevent membrane lipid peroxidation and played a defensive role (Zhang et al. 1993). Therefore, the increase of isozyme patterns of sugarcane varieties under low temperatures can be used as one of the indexes of cold resistance of sugarcane.

(2) Proline content

Due to the self-protection response of plants, the content of free proline in plant tissues will change significantly when they are subjected to various stresses. Therefore, proline is often used as a physiological index of stress resistance. The results showed that the proline content in the cultivars was higher than that in the other varieties. The results showed that the free proline content in wild sugarcane species increased gradually in autumn and winter under natural environmental conditions. There was a strong positive correlation between cold resistance and proline content. Therefore, it is feasible to use proline as a biochemical index to study cold resistance in sugarcane. It has been reported that proline content changes as the identification index of cold resistance in sugarcane, Brassica napus, cucumber, eggplant, and other plants (Ding, 2001; Peng et al., 2004).

(3) Plasma membrane permeability

Under the action of low-temperature stress, the cell membrane will be damaged to a certain extent, and the electrolyte in the cell will leak out. Therefore, in the study of sugarcane stress resistance, the electrical conductivity method can determine the stress resistance of plants. The research results of Chen (1990), Chen et al. (1996), and Zhu (1995) showed that the conductivity method could be used to distinguish the difference in cold resistance of different sugarcane varieties. In a specific range of low-temperature treatment and time, the electrical conductivity was significantly negatively correlated with the cold resistance of sugarcane varieties. The differences were utterly consistent with the cold resistance performance of varieties. The results showed that the electric conductivity decreased with the elevation of the original growing place and decreased with the increase of the latitude of the original growing place. In a similar latitude range, the factor determining the electric conductivity was the altitude, and in a similar altitude range, the factor determining the electric conductivity was latitude (Dai, 1989).

(4) Chlorophyll content

After low-temperature damage, the chlorophyll content in plant leaves will be destroyed. With the aggravation of low temperature, the chlorophyll content changes significantly, showing a downward trend. After freezing, the chlorophyll content of sugarcane varieties was different, and the decrease of chlorophyll content was consistent with the cold resistance of sugarcane varieties (Chen et al., 1996).

5.3.3.2 Effect on photosynthesis of sugarcane

The effect of low-temperature stress on the photosynthesis of sugarcane is multifaceted. It can directly induce the damage of photosynthetic structure and affect photosynthetic electron transport, photosynthetic phosphorylation, and dark reaction-related enzymes. Under low-temperature stress, the primary light energy conversion efficiency and potential photosynthetic activity of sugarcane PS II were inhibited. The degree of inhibition was that the varieties with poor cold resistance were significantly greater than those with solid cold resistance. At the same time, low-temperature stress also directly affected the carbon assimilation of photosynthesis. The results showed that: under low-temperature stress, chlorophyll a fluorescence induction kinetics curve of sugarcane leaves changed significantly, variable fluorescence maximum fluorescence, variable fluorescence initial fluorescence, DCPIP photo reductive activity, quantum yield and fluorescence decline rate decreased, fluorescence quenching slowed down, photochemical quenching coefficient and non-photochemical quenching coefficient decreased (Zhang et al., 1999).

5.3.3.3 Effect on sugarcane production

The impact of low-temperature disasters on sugarcane production is related to many factors, such as the type of low-temperature disaster, the time of disaster occurrence, the intensity of disaster occurrence, and the weather change after the disaster. Generally speaking, a low-temperature disaster not only causes the decline of sugarcane yield, sucrose content, juice quality, and sugar yield rate, but also freezes the remaining sugarcane buds, resulting in the decrease of seedling emergence rate and ratoon number, and even affect the growth of sugarcane in the next year. No matter what type of low-temperature disaster, as long as its intensity is significant and early, it will significantly impact sugarcane production.

Sugarcane yield: After the sugarcane is damaged by freezing, if the growth point dies, the sugarcane plant stops growing and can only sprout lateral buds; after the leaf is damaged by freezing, photosynthesis cannot be carried out, affecting the synthesis of organic matter; the sugarcane damaged by freezing will cause hollow cane, the light consumption rate can reach 10%; the whole cane damaged sugarcane will deteriorate and reduce the weight.

Sugarcane seedlings: When sugarcane stems are seriously frosted, the growth points and lateral buds die, significantly affecting the quantity and quality of sugarcane seedlings.

Ratoon germination: When the sugarcane plant is damaged by freezing, the lateral buds at the base of the sugarcane stem will die, which will lead to the decrease of ratoon number, the delay of sprouting, and poor growth, which will affect the yield of ratoon cane in the coming year.

Sugar content and quality: After sugarcane is damaged by freezing, the loss of sucrose can reach 5 %–10 %, and the reducing sugar composition will increase, the acidity and colloid of sugarcane juice will increase, which will reduce the quality of sugarcane juice, seriously affect the production of sucrose and reduce the sugar yield.

5.3.4 Breeding and evaluation of cold-resistant sugarcane varieties

The way crops complete the growth cycle before the stress is free from damage is called the "escape" mechanism, which is usually harmless to yield and is a critical mechanism for the cold resistance of sugarcane. Low temperature and cold damage seriously affect the yield and quality of sugarcane. Therefore, it is appropriate to select the cold-resistant varieties with early maturity and high sugar content, early growth and rapid development, and easy storage in frost areas.

5.3.4.1 Identification of cold resistance of sugarcane varieties

For a long time, researchers have devoted themselves to the identification of cold-resistant sugarcane varieties. The evaluation of cold resistance mainly focused on the stalk, leaf, bud, and ratooning ability after freezing injury. The physiological and biochemical metabolism and biological characteristics of seedlings and leaves of different varieties under artificial temperature control were compared. Li et al. (1998) identified 'CMT88-68' 'CMT70-611' and 'CMT86-05' with a germination rate of seed stem at low temperature and cell membrane permeability of detached leaves after freezing as identification indexes. Huang et al. (2002) and Lu et al. (2010) evaluated the cold resistance of different sugarcane varieties by analyzing physiological and biochemical indexes and selected several varieties with solid cold resistance. Wei et al. (2009) and Yang et al. (2011) compared the cold resistance of different sugarcane varieties in terms of agronomic traits and selected several varieties with solid cold resistance and simple and easy evaluation indexes of cold resistance (Table 5-1).

Table 5-1 Identification methods and effects of sugarcane cold resistance

Test index	Identification conditions	Appraisal effect	Difficulty of implementation	Method evaluation
Germination of stem at low temperature	9–15 °C	Obvious	Easy	Feasible
The germination rate of freezing stem	–2 °C, 6–12 h	Obvious	Easy	Feasible
Seedling growth after low-temperature treatment	0 °C, 24–48 h	Some obvious	Easy	Feasible
Irrigation culture with ice water mixture	Irrigation for 1 month	Mid	Easy	Restricted
Determination of root activity	–2 °C, 24 h	obvious	More difficult	Restricted
Determination of proline content	–2 °C, 24 h	obvious	More difficult	Restricted
Chlorophyll degradation	–2–4 °C, 8–15 h	Some obvious	Mid	Restricted
Determination of cell membrane permeability	–5 – –3 °C, 7–16 h	obvious	Mid	Feasible
Number of perennial roots	After natural frost in the field	obvious	Some easy	Feasible

Note: "restricted" means it is applicable to identify varieties or clones with a significant difference between cold resistance performance and standard variety.

5.3.4.2 Cold resistance breeding of sugarcane

Sugarcane is a thermophilic crop, which needs appropriate temperature and humidity to grow normally. With the introduction of *S.spontaneum* resistance genes, sugarcane planting has expanded from tropical to subtropical and temperate regions. The leading sugarcane-producing areas in China are mainly distributed in subtropical regions. Low temperature and cold damage seriously affect the yield and quality of sugarcane. Especially in recent years, climate change is abnormal, extremely low temperature occurs frequently, and sugarcane is seriously affected by cold disasters, which causes significant economic losses to sugarcane farmers and sugar enterprises. Therefore, the selection of sugarcane cold-resistant varieties and the screening of vigorous cold-resistant varieties in the sugarcane area are significant to the stable development of sugarcane production, the sugar industry, and other sugar enterprises.

In recent years, significant progress has been made in the breeding of new cold-resistant sugarcane varieties. The introduction of new genetic genes can increase the frequency of cold-resistant lines and enhance the cold resistance of sugarcane lines. Tai (1993) reported that frost could damage the stems and leaves and reduce the perennial root yield, while the cold-tolerant varieties can alleviate the damage. Brandes (1939) reported the possibility of breeding cold-resistant varieties using the materials of secretory hand dense. Irvine (1967) also reported that the cold tolerance of other materials and several related genera of sugarcane was more substantial than cultivated lines, which could be used. Arcncanx studies have shown that cold resistance can be inherited. Yang (1996) further studied the relationship between cold resistance and Ratooning in sugarcane and found a significant positive correlation between them. The proportion of the close blood determined the ratooning ability. The cold resistance is mainly controlled by the general combining ability of the male parent.

In contrast, the ratooning ability is affected by the parents' general combining ability and special combining ability. The cold resistance and ratoon of sugarcane varieties can be improved by parent selection and combination selection. Wu (1996) studied the cold resistance of 67 single lines selected from 24 different cross combinations. The results showed that the parents had solid cold resistance, and their offspring had a higher frequency of cold-resistant varieties and better cold resistance. The parents with the worst cold resistance were 'CP57-614' and 'CCZ4', and the parents with the strongest cold resistance were 'CYC71-374' and 'Xuan 30'. Chen et al. (2012) investigated the agronomic traits of introduced sugarcane varieties under natural low-temperature conditions with 'ROC22' as the control. They comprehensively classified the cold resistance of various varieties using the membership function value method and analyzed and compared the physiological and biochemical indexes of various cold-resistant varieties under artificial low-temperature conditions. The results showed that the cold resistance of 'Taitang 88-99' 'CP84-1198' and 'CYZ99-601' was strong; 'CGZ18' 'CL-2003'. 'YuanLin9' 'Taitang 98-0432' 'CFN15' 'Taitang98-1626' and 'CYT96-86' belong to moderate cold resistance varieties, while 'YuanLin 6' and 'YuanLin 8' have weak cold resistance.

Low temperature and freezing injury is an essential condition for cold resistance breeding of sugarcane. In the area where low temperature and freezing damage occur frequently, non-cold-resistant materials will be gradually eliminated. The severe freezing injury in the Sichuan sugarcane region has periodicity characteristics, which creates a good condition for breeding cold-resistant sugarcane varieties. The Sugarcane Research Institute, Yunnan Academy of Agricultural Sciences (YSRI), and the Sugar Crops Research Institute of Sichuan Academy of Plant Engineering have established a cooperative relationship with cold-resistant sugarcane breeding in Zizhong, Sichuan Province. The cold resistance of the excellent sugarcane varieties bred in recent years was evaluated, and the excellent cold-resistant sugarcane varieties such as 'CYZ03-194' 'CYZ05-51' and 'CYZ05-39' were selected (Wu et al. 2017). The yield, disease resistance, brix, cold resistance, and sprouting ability were better than 'ROC22'. Parents strictly control cold resistance. Usually, only parents with high cold resistance can produce offspring with solid cold resistance.

Therefore, it is expected to produce offspring with solid cold resistance by using the cold-resistant sugarcane parents such as 'CYZ03-194' 'CYZ05-51' 'CYZ99-601' 'CYT96-86' 'CYC71-374' and 'Xuan30' as parents, which is expected to produce offspring with strong cold resistance, which is expected to play an essential role in sugarcane cold resistance breeding in China. It is of positive significance to promoting sugarcane cold resistance breeding in China.

It is one of the most effective measures to reduce the cold resistance of sugarcane. However, the cold resistance of different sugarcane genotypes is quite different, and the cold resistance of sugarcane needs to be further studied. In addition to the physiological, biochemical, and molecular level research on the mechanism of low-temperature damage, combined with the morphological characteristics of sugarcane plants, the identification methods of cold resistance of sugarcane varieties were further improved. The molecular biology and genetic engineering methods were applied to sugarcane resistance. Cold-resistant sugarcane varieties should be cultivated popularized in low-temperature damaged areas.

References

BRANDES E W, 1939. Three generations of cold resistant sugarcane. Sugar Bull, 18 (4):3-5.

CHEN C J, WEI H W, WU J M, et al., 2012. Evaluation on cold resistance of sugarcane varieties newly introduced to Guangxi. Journal of Southern Agricultural, 43 (6):744-748.

CHEN N W, 1990. The identification of cold resistance by means of electrical conductivity in sugarcane. Sugarcane and Canesugar (6):14-20.

CHEN N W, YANG R Z, WU C W, et al., 1996. Studies on the technic of identification for freeze resistance in sugarcane. Sugarcane and Canesugar (4):1-9.

CHEN S Y, CHEN R K, CHEN Q F, et al., 1994. The protective effect of free radical scavenger and drought resistance in sugarcane, Acta Agronomica Sinica, 20 (2):149-155.

CHEN Y Q, DENG Z H, GUO C F, et al., 2007. Drought Resistant Evaluations of Commonly Used Parents and Their Derived Varieties. Scientia Agricultura Sinica, 40 (6):1108-1117.

DAI X Y, 1989. Study on cold resistance of *Saccharum Spontaneum* L. means of Electrolyte leakage. Journal of Yunnan Agricultural University, 4 (3):15-22.

DING C, 2001. Application of electrical conductivity and free proline in identification of cold resistance of wild sugarcane germplasm resources. Kunming:Yunnan Agricultural University, master's thesis.

EVANS H, 1935. Some aspects of the problem of drought resistance in sugarcane. Proceedings of International Society of Sugar Cane Technology (6):802-808.

EVANS H, 1937. A preliminary study of root characters as affecting drought resistance in two sugarcane varieties and their seeding. Sugar cane research station, Mauritius Bulletin (14):56-62.

FU C, LIU S M, HU H X, et al., 2009.Evaluation of backcross selection of sugarcane *arundinacea* interspecific distant hybrid Yacheng 01-87. Journal of Crops (1):35-39.

GAO S J, LUO J, CHEN R K, et al., 2002. Photosynthetic physiological indexes of the drought resistance of sugarcane and its comprehensive evaluation. Acta Agronomica Sinica, 28 (1):94-98.

GAO SJ, LUO J, ZHANG H, et al., 2006. Physiological and biochemical indexes of drought resistance of sugarcane (*Saccharum spp.*). Acta Zoologica Sinica, 17 (6):1051-1054.

HUANG Y Z, XU J Y, CHEN C J, et al., 2002. Comparative test on drought resistance and frost resistance of several new sugarcane varieties. Agricultural Biology of Guangxi, 21 (2):101-104.

IRVINE J E, 1967. Testing sugarcane varieties for cold tolerance in Louisiana. Proceedings of 12th Congress of international sugar cane technologist. Puerto Rico, 569-574.

JIN Y F, TAO L A, YANG L H, et al., 2002. A preliminary report on the drought Resistance of sugarcane F_3 cloneswith Yunnan wild breed. Sugarcane, 9 (1):19-21.

LIANG L Q, TAN Y M, ZENG J Q, et al., 1995. A studies on relation between ecological characteristics, proline content, membrane permeability and drought tolerance in sugarcane leaves. Sugarcane, 2 (4):14-19.

LI M Z, 1998. The cold resistance of sugarcane and its measures. Sugar Crops of China (2):42-45.

LI R M, YANG K Z, LIN Y X, et al., 1998. Identification of cold resistance of several self-bred sugarcane varieties. Sugarcane, 5 (4):5-10.

LI Y R, 2010. Modern sugarcane science. Beijing:China Agricultural Press (in Chinese).

LU G Y, WEI H, JIANG M M, et al., 2010. Effects of soaking sugarcane in ice-water on physiological and biochemical characteristics. Guangxi Agricultural Science, 41 (2):113-116.

LUO J, LIN Y Q, 2000. Response of chlorophyll a fluorescence induction kinetics parameters of sugarcane leaves exposed to drought stress. Sugarcane and Canesugar (2):15-20.

LUO M Z, LIU Z F, LIANG J N, et al., 2005. The relationship between drought resistance of sugarcane and some physiological biochemical properties of leaves. Subtropical Agriculture Research, 1 (1):14-16.

PAN S M, CHEN Y Q, WU J S, et al., 2006. Screening and evaluation of the drought-resistant germplasm in sugarcane. Acta Agricultural University Jiangxiensis, 28 (6):838-444.

PENG H L, SU Z X, 2004. Effects of low temperature stress on physiological and biochemical indexes of cold resistance of Pittosporum seedlings. Journal of Hanzhong Teachers College (Natural Science), 22 (2):50-53.

TAI P Y P, 1993. Low temperature preservation of F1 pollen in crosses between noble or commercial sugarcane and *Saccharum spontaneum*. Sugarcane, 5:8-11.

TAN Y M, 1988. A study on membrane fatty acids and permeability in sugarcane leaves related to drought resistance. Journal of Fujian Agricultural University, 17 (3):211-215.

WEI H, LU G Y, HAN S J, et al., 2009. Effects of seed cane soaking in ice water on seedling emergence rate and seedling quality of different sugarcane varieties. Guangxi Agricultural Sciences, 40 (6):629-632.

WU C W, 1996. The role of low temperature and freezing injury in sugarcane breeding. Fujian Sugarcane (1):26-31.

WU C W, LIU J Y, YANG K, et al., 2011. Evaluation of Ratooning and drought resistance of Hybrid Progenies of Yunrui innovative parents. 2011 Sugarcane Industry Development Forum and 14th Symposium of Sugarcane Professional Committee of Chinese Crop Society:1-10.

WU C W, ZHAO J, XIAO Y, et al., 2017. Evaluating on cold resistance and performance of sugarcane clones in the north marginal cane-growing region. Sugar Crops of China, 39 (6):8-12.

XU Y S, 1986. Physiological response of callus of sugarcane to water stress and their drought resistance. Nanning: Guangxi Agricultural University.

YANG L H, TAO L A, JIN Y F, et al., 2008. Drought-resistance Heredity analysis of Yunnan wild sugarcane. Sugar Crops of China (4):10-13.

YANG R Z, 1996. The preliminary study on cold-resistance and ratooning of ability of sugarcane. Sugarcane and Canesugar (6):13-17.

YANG R Z, 1999. A preliminary analysis on quality deterioration of several sugarcane varieties under natural frozen injury. Sugarcane, 6 (2):1-5.

YANG R Z, LI Y R, WANG W Z, et al., 2011. Evaluation on cold tolerance of sugarcane under drought and frost conditions. Southwest China Journal of Agricultural Sciences, 24 (1):52-57.

YE Y P, LI Y R, TANG J, et al., 2003, Studies on some physiological indices for drought-resistance in two sugarcane varieties under drought and irrigation conditions. Sugar Crops of China (1):1-5.

ZENG H Z, ZHENG C M, 2003. Physiological test of drought resistance and PCR amplification of BADH Gene in sugarcane. Chinese Journal of Tropical Crops, 24 (1):55-58.

ZHANG C L, ZHOU J Y, WU X Y, et al., 2000. Frost damage to sugarcane yield and sugar content in Menghai County. Sugarcane, 7 (4):52-53.

ZHANG M Q, CHEN R K, 1993. Studies on the cold resistance of sugarcane II. Changes of the enzymes in Seedling Leaf under low temperature treatment of seed cane. Journal of Fujian Agricultural College (Natural Science Edition), 22 (1):23-27.

ZHANG M Q, CHEN R K, LV J L, et al., 1999, Effects of low temperature stress on Chlorophyll a fluorescence induction kinetics in the seedling of sugarcane. Journal of Fujian Agricultural University, 28 (1):1-7.

ZHANG Y B, 2011. Development technology of sugarcane industry in China. Beijing:China Agricultural Press (in Chinese).

ZHONG X Q, LIN L C, 2002. The relationship between the drought-resistance of sugarcane varieties and physiological characteristics. Journal of Foshan University (Natural Science Edition), 20 (3):59-62.

ZHU Q Z, 1995. Identification of cold resistance of sugarcane. Agricultural Sciences of Guangxi (6):264-265.

6 Variety Techniques and New Varieties in Sugarcane

6.1 Naming of Sugarcane Varieties

Sugar-producing countries did not carry out large-scale sexual cross-breeding work at the end of the 19 th century. The number of varieties was less, and the varieties were simple, which was not easy to be confused. After self-breeding, sugarcane varieties in various countries have gradually become more complex and diverse, so naming has become an issue that sugarcane breeders must consider in various countries. If there is no unified system for naming sugarcane varieties globally, the world's chaos will be unimaginable.

There are two main methods for naming sugarcane varieties in the world.

Continuous method: After many years of testing, screening, or appraisal (examination) of varieties, the continuous method is often used. The symbols (names) of the varieties are continuous, such as 'Co290' 'Co409' 'Q170' 'Q171' 'POJ2878' 'F134' 'F135' 'C 1' 'CGT2' 'CCZ1' 'CCZ2' etc.

Discontinuous method: Numbering according to the number of seedlings per year, but the number of good seedlings is not continuous, the number in front of the variety is the year, and the number behind is the number of seedlings, such as 'CYZ03-194' 'CYZ06-407' 'CGT94-119' 'CFN91-3621' 'CYT 93-159' 'CCT 61-408' 'CP36-105' 'CP72-1210' etc., but in some western countries, the number in front is the seedling selection number, and the number behind is the year, such as the variety C266-70 bred in Cuba.

In addition, there are individual organizations that use mythological characters to name them, such as the Australian Colonial Sugar Refining Company (CSR), whose varieties are named 'Trojan' 'Pindar' etc. individual countries use islands or place names, such as Mali and Mama in Fiji.

Due to the increase of breeding units, the rapid increase of variety symbols, and the possibility of repeated variety names, the International Sugarcane Technicians Association Germplasm Committee has formulated some principles to standardize the name sugarcane varieties. So far, the committee has only recognized the symbols listed by Daniels and the names later approved by the Chairman of the Committee (lately D. J. Heinz) (Peng, 1990). Therefore, anyone who wants to adopt a new symbol name should first obtain the chairman's approval of the International Sugarcane Technicians Association Germplasm Committee. It will ensure that the selected symbol has not been used in the past and conforms to the variety set by the committee. The naming standard is recognized by the International Sugarcane Technicians Association and accepted by sugarcane research institutes worldwide.

6.2 Symbols of Sugarcane Varieties and Their Breeding Organizations

More than 90 cane sugar-producing countries or regions worldwide cover America, Asia, Africa, Europe, and Oceania. America is the world's largest cane sugar-producing area, and 35 countries or regions produce cane sugar. Asia is the second-largest cane sugar-producing region in the world, with 16 countries producing sugar. Africa is the third-largest cane sugar-producing region in the world, with 35 countries producing sugar. Three countries produce cane sugar in Oceania and at least 2 countries in Europe (Table 6-1). Generally speaking, each cane sugar-producing country has sugarcane breeding organizations. Some cane sugar-producing countries have breeding organizations in their internal provinces (states) and large enterprises. The symbol and breeding units of sugarcane varieties in various countries are as follows.

Table 6-1 Statistics of cane sugar-producing countries in the world

No.	Country	continent	No	Country	continent	No	Country	continent
1	China	Asia	4	Pakistan	Asia	7	Viet Nam	Asia
2	India	Asia	5	Philippines	Asia	8	Myanmar	Asia
3	Thailand	Asia	6	Indonesia	Asia	9	Bangladesh	Asia

continued

No.	Country	continent	No	Country	continent	No	Country	continent
10	Iran	Asia	38	French Guiana	America	66	Cameroon	Africa
11	Sri Lanka	Asia	39	French west indies	America	67	Reunion	Africa
12	Malaysia	Asia	40	Guadeloupe	America	68	Malawi	Africa
13	Nepal	Asia	41	Haiti	America	69	Sierra Leone	Africa
14	Cambodia	Asia	42	Canada	America	70	Senegal	Africa
15	Iraq	Asia	43	Martinique	America	71	Uganda	Africa
16	Japan	Asia	44	St base	America	72	Zambia	Africa
17	Brazil	America	45	Trinidad	America	73	Chad	Africa
18	Mexico	America	46	Uruguay	America	74	Somalia	Africa
19	Columbia	America	47	Jamaica	America	75	Rwanda	Africa
20	USAs	America	48	Barbados	America	76	Angola	Africa
21	Argentina	America	49	Antigua	America	77	Congo	Africa
22	Guatemala	America	50	Grenada	America	78	Zaire	Africa
23	Cuba	America	51	Surinam	America	79	Ghana	Africa
24	Venezuela	America	52	South Africa	Africa	80	Nigeria	Africa
25	Peru	America	53	Egypt	Africa	81	Gabon	Africa
26	Ecuador	America	54	Sudan	Africa	82	Cotedlvoire	Africa
27	Bolivia	America	55	Swaziland	Africa	83	Liberia	Africa
28	Dominica	America	56	Kenya	Africa	84	Guinea	Africa
29	Salvador	America	57	Mauritius	Africa	85	Niger	Africa
30	Honduras	America	58	Zimbabwe	Africa	86	Morocco	Africa
31	Nicaragua	America	59	Tanzania	Africa	87	Australia	Oceania
32	Costa Rica	America	60	Madagascar	Africa	88	Fiji islands	Oceania
33	Paraguay	America	61	Mozambique	Africa	89	French Polynesia	Oceania
34	Guyana	America	62	Mali	Africa	90	Portugal	Europe
35	Panama	America	63	Ethiopia	Africa	91	Spain	Europe
36	Puerto Rico	America	64	Burkina Faso	Africa			
37	Belize	America	65	Burundi	Africa			

6.2.1 Symbols of domestic sugarcane varieties and their breeding organizations

6.2.1.1 CGX (Guangxi)

New sugarcane varieties are named in Guangxi Zhuang Autonomous Region, China.

CGT (Guitang), Sugarcane Research Institute of Guangxi Zhuang Autonomous Region, was later renamed Sugarcane Research Institute of Guangxi Academy of Agricultural Sciences.

CLC (Liucheng), Sugarcane Research Center of Liucheng County, Guangxi.

6.2.1.2 CYN (Yunnan)

New sugarcane varieties are named in Yunnan Province, China.

CYZ (Yunzhe), Sugarcane Research Institute of Yunnan Academy of Agricultural Sciences.

CYR (Yunrui), Ruili Breeding Station of Sugarcane Research Institute, Yunnan Academy of Agricultural Sciences.

CDZ (Dezhe), Sugarcane Research Institute of Dehong Prefecture, Yunnan.

CLH (Lianghe), Sugar Production Office of Lianghe County, Dehong Prefecture, Yunnan Province.

6.2.1.3 CGD (Guangdong)

New sugarcane varieties are named in Guangdong Province, China.

CYT (YueTang), Guangdong Sugar Cane Sugar Research Institute, was later renamed the Cane Sugar Research Institute of the Ministry of Light Industry, and now Guangzhou Sugar Cane Sugar Research Institute.

CHN (Huanan), South China Institute of Agricultural Sciences (now Guangdong Academy of Agricultural Sciences).

CYN (Yuenong), Guangdong Agricultural Science and Economic Crop Research Institute.

CHZ (Haizhe), Hainan sugarcane breeding Station.

CYC (Yacheng), Lines or parents bred in Hainan Sugarcane Breeding Farm.

CZZ (Zhanzhe), Zhanjiang Sugarcane Test Station.

CST (Shuntang), Shunde Sugarcane Test Field.

6.2.1.4 Fujian (CFJ)

New sugarcane varieties are named in Fujian Province, China.

CFN (Funong), Sugarcane Research Institute of Fujian Agriculture and Forestry University.

CMT (MinTang), Sugarcane Research Institute of Fujian Academy of Agricultural Sciences (later renamed Sugarcane Research Institute of Fujian Academy of Agricultural Sciences).

CMN (Minnong), Fujian Agricultural College.

6.2.1.5 Taiwan (CTW)

New sugarcane varieties are named in Taiwan Province, China.

EG, Former Taiwan Saltwater Port Sugar Co., Ltd.

F, Taiwan Sugar Research Institute, China.

Fm, Taiwan Sugar Research Institute, Taiwan seedling male parent.

Fp, Taiwan Sugar Research Institute, Taiwanese parent.

PT, Sugarcane Research Institute of Pingtung, Taiwan Province.

ROC, Taiwan Sugarcane Industry Research Institute, China.

ROC (New Taiwan Sugar), A variety newly bred by the Taiwan Sugar Research Institute.

TA, Excellent varieties selected from TM.

Tm, Former Taiwan Sugar Research Institute Co., Ltd.

6.2.1.6 Sichuan (CSC)

New sugarcane varieties are named in Sichuan Province, China.

CCZ (ChuanZhe), Variety by the Sichuan Sugar Industry Research Institute, was later renamed the Sichuan Academy of Botanical Engineering.

CCT (Chuantang), The clones and materials by the Sichuan Sugar Industry Research Institute, was renamed the Sichuan Academy of Botanical Engineering lately.

CCN (Chuanning), Ningnan Test Station of Sichuan Sugar Research Institute.

CTC (Tiancheng), Sugarcane Experimental Farm of Neijiang City, Sichuan Province (later renamed as the Neijiang City Agricultural Science Institute, now Neijiang City Academy of Agricultural Sciences, Sichuan Province).

CLZ (Liangzhe), Subtropical Crops Research Institute of Liangshan Prefecture, Sichuan Province.

6.2.1.7 Jiangxi (CJX)

New sugarcane varieties are named in Jiangxi Province, China.

CGN (Gannan), The clones and materials by Sugarcane Scientific Research Institute of Ganzhou, Jiangxi.

CGZ (Ganzhe), Variety by Sugarcane Scientific Research Institute of Ganzhou, Jiangxi.

6.2.2 Symbols of foreign sugarcane varieties and their breeding organizations

A, Argentina.

AB, Antigua Island.

AH, Java.

B, The Sugarcane Breeding Station in the West Indies Center of Barbados.

Ba, Barbados.

BBZ, Belize Sugar Research Institute.

Bc, Variety bred in the later stages of the Barbados breeding farm.

BF, Hybrid in Barbados, varieties bred by Pakistan (Faisalabad, Punjab) Sugarcane Research Institute.

Bh, Hybrid in Barbados.

BJ, Hybrid in Barbados, varieties cultivated at Jamaica Sugar Research Institute.

BL, Hybrid in Barbados, varieties cultivated in Lyallpur, Pakistan.

BO, Variety bred in Bihar and Orissa, India.

BR, Hybrid in Barbados, varieties cultivated by Romana in the Dominican Republic.

BT, Hybrid in Barbados, varieties cultivated by Caroni Research Station in Trinidad and Tobago.

C, Agricultural Experiment Station, Santigo de las Vegas, Cuba.

C.B. Varieties bred in the Breeding Farm of Campos University, Brazil.

CAC (Cane of Agricultural College), Varieties cultivated by the College of Agriculture of the University of the Philippines.

CaSe, Hybrid in India Coimbatore, varieties bred at Seorahi Sugarcane Research Institute.

CB, Varieties bred in the breeding farm of the University of Campos, Brazil.

CC (College Cane), Varieties cultivated by the College of Agriculture, University of the Philippines.

CCCP, Hybrid in Canal Point, varieties bred by Guatemala Sugar Research Center.

CG, Hybrid in Guatemala, varieties bred by Guatemala Sugar Research Center.

CGM, Hybrid in Mexican, variety bred by Guatemala Sucrose Research Center.

CGSP, Hybrid in Brazil (Copersucar), varieties bred at the Guatemala Sugar Research Center.

CH, Cuban Hydrid.

CIMCA, Santa Cruz la Sierra.

CL (Clewiston, Florida U.S. Sugar Corporation), Varieties bred at the Sugarcane Research Station in Clewiston, Florida.

Co. (Coimbatore, India), Coimbatore Sugarcane Breeding Station in India.

CoA, Hybrid in India Coimbatore, varieties bred at Anakapelle (AP) Sugarcane Research Institute.

CoC, Hybrid in India Coimbatore, varieties bred at the Sugarcane Research Institute of Cuddalore (TN).

CoH, Hybrid in India Coimbatore, varieties bred at Haryana Sugarcane Research

Institute.

CoJ (Coimbatore-Jullundur, East Punjab), Hybrid in Indian Coimbatore, varieties cultivated at Punjab University in Jalandhar, India.

CoK (Coimbatore-Karnal), Hybrid in India Coimbatore varieties bred at the Kanal Sugarcane Test Farm in northern India.

CoL (Coimbatore, Lucknow, India), Hybrid in Indian Coimbatore, bred in Lyallpur before India was separated from Pakistan.

Colk, Hybrid in India Coimbatore, varieties bred at Lucknow Sugarcane Test Farm in India.

CoM, Hybrid in India Coimbatore, varieties bred at Maharashtra Sugarcane Research Institute, India.

CoN, Hybrid in India Coimbatore, varieties bred at Navsari Sugarcane Research Institute, India.

CoP, Hybrid in India Coimbatore, varieties bred at Pasa Sugarcane Research Institute, India.

CoPant, Hybrid in India Coimbatore, varieties bred at the Pantnagar Sugarcane Research Institute in India.

CoR (Coimbatore-Risalewala), Hybrid in India Coimbatore, varieties bred in Risarewala.

CoS (Coimbatore-Shahjahanpur), Hybrid in Indian Coimbatore, bred at the Shahjahanpur Sugarcane Test Site in northern India.

CoSe (Coimbatore-Seorahi), Hybrid in India Coimbatore, varieties bred in Seorahi, India.

CoT, Hybrid in India Coimbatore, varieties bred at Tirupati Sugarcane Research Institute.

CoTL, Hybrid in India Coimbatore, varieties bred at Thiruvalla Sugarcane Research Institute.

CoV, Hybrid in India Coimbatore, varieties bred at Vuyyuru Sugarcane Research

Institute.

CP (Canal Point, Florida, U.S.A.), Hybrid in Canal Point, Florida, USA, varieties bred at Canal Point, Florida, USA, and the Sugarcane Breeding station of the Homer Department of Agriculture, Louisiana.

CP, Variety bred by Pakistan NWFP Sugarcane Research Institute.

CPM Mardan, Pakistan, varieties bred by NWFP Sugarcane Research Institute.

CR (Dom.Rep., Central Romana) Romana, the center of the Dominican Republic.

D (Demerara), Variety bred by Guyana Demerara Sugar Co., Ltd.

DB (Demerara-Barbados), Hybrid seed in Barbados , vareity bred at Guyana Demerara-Barbados.

DI (Dimak Idjo, Java), Java Tammu Izo.

E, Egypt.

EPC (EstaCon Experimental, Palmira, Columbia), Sugarcane Test Field in Palira, Colombia.

EA (East African Sugar Cane Breeding Station, Kenya), Kenya, Sugarcane Breeding Station in East African.

EK (Edward Karthaus), The seedlings cultivated by Edward Karthaus, Java.

F (Florida), Sugarcane Breeding Station Florida, USA.

FAM, Crossed and bred in Tucuman, National Institute of Agricultural Technology in Argentina.

FC, Fajardo Central Station in Puerto Rico.

FR, Hybrid in Guadeloupe, Ivory Coast, and variety selected by CIRAD-CA, Agricultural Development Research Center in French.

G, Giza in Egypt.

GC, Puerto Rico Guanica Sugar Factory.

GPB, Variety bred in GPBPD, Malaysia.

GT, Hybrid from Taiwan, China, varieties bred by SCRI in Egypt.

GZ, Java Series.

H, Hawaii Agricultural Research Center, USA.

Hocp, Hybrid in Canal Point of Florida, USA, and variety bred at the Sugarcane Breeding Station of the Department of Agriculture, Homer, Louisiana.

HJ, Hybrid in Hawaiian, varieties cultivated in Jamaica.

HM, Hebbal, Mysqre in India.

HQ, Queensland Hambledom Sugar Company in Australia.

HVA, Java Handal Vereeming Amstasdame in Indonesia.

I, Ishurdi, Bangladesh.

IAC, São Paulo Campinos Agricultural Institute in Brazil.

IACSP, Institute of Agronomy of Campinos São Paulo in Brazil.

IANE, Northeast Agricultural Research Institute in Brazil.

ICA, Agricultural Research Institute in Colombian.

Isd, Sugarcane Research Institute in Bangladesh.

J, Sugar Research Institute in Jamaica.

JAR, H.G.Sorensen Central Jaronu , Cuba.

Ja, Cuba.

K, Cane and Sugar Industry Promotion Center, Region one (Kanchanaburi Province), Thailand.

KK, Field Crops Research Center of Khon Kaen Province in Thailand.

KHS, Hybrid on Coimbatore, India, and varieties bred in Karmataka sugarcane hybrid farm.

KEN, Kenya Sugar Research Foundation on Kisumu, Kenya.

Kn, Hybrid in Sudan, varieties bred in Sudan Kenana Sugar Co., Ltd.

L, Lyallpur in Pakistan.

L, Varieties cultivated by Louisiana State University, USA.

La, Peru.

Laica, Varieties bred by Sugar Research Institute, Costa Rica.

LCP, Hybrid in Canal Point, Florida, USA, and variety bred by Louisiana State

University.

LF, Hybrid, and varieties bred from Fiji Sugar Company in Lautoka.

LHO, Crossbreed on Homer, selecting on Louisiana, breeding on Louisiana State University.

Lk, Hybrid on Coimbatore India, varieties bred by Lucknow Sugarcane Research Institute.

M, Sugar Research Institute in Mauritius.

M, Hybrid on Coimbatore, India, varieties bred at Maharashtra Sugarcane Research Institute.

Mayari, Sugarcane test site on Mayari, Cuba.

Mer, Meridian in the United States.

Mex, National Sugar and Liquor Research Institute in Mexican.

MexBz, Hybrid in Tapachula, Belize Sugar Research Institute.

ML (Media Luna), Sugarcane test field in Meridiana, Cuba.

MP, Mauritius Perromat.

MPR, Proving Ground in Mayaguez, Puerto Rico.

MPT, Mitr Phol Sugarcane Research Centre, Phukieo, Chaiyaphum, Thailand.

MQ, Mackay Sugarcane Test Field in Queensland, Australia.

My, Cuba.

MZC, Colombia Mayaguez Sugar Mill.

N, South African Sugar Association in Natal, South Africa.

NA, Experimental Base of Northern Argentina.

NCo, Hybrid on Coimbatore, India, and variety cultivated in Natal.

NG, New Guinea.

NH, New Hebrides.

Ni (Nippon), Agriculture and Forestry Experiment Station on Okinawa Prefecture, Japan.

NiF, Hybrid on Taiwanese, varieties bred of Japan in Okinawa, South Africa.

NiN, Hybrid on Natal, varieties bred of Japan in Okinawa, South Africa.

NM, Natal, South Africa.

P, Peru.

PB, Plant Breeding Station, Philippines.

PGM, Mexican Ajiao seeds, bred in Guatemala.

Phil, Philippine Sugar Authority.

PI, Parry Limited, Bangalore, India.

PM, Potrero Mexico.

PO, Hybrid in Argus, Brazilian, and varieties bred in Sao Paulo.

POJ (Proefstation Oast Java), East Java Test Station in Indonesia.

PR (Rio Prdra Jusula Experiment Station), Puerto Rico Rio Pedras Isolation Test Station.

PRN, Luzon hybrid, bred by the Philippine Sugar Research Institute.

PS (Pasuruan, Java), Bred in East Java (Pasuruan) by Indonesia Sugar Research Center.

PSA, Sugar Association in the Philippines.

PSBM, Hybridized in Pasuruan, East Java, and Indonesia Sugar Research Center bred in bangamayang.

PSCO, Hybridized in Pasuruan, East Java, and Indonesia Sugar Research Center bred in Comal.

PSJT, Hybridized in Pasuruan, East Java, and Indonesia Sugar Research Center bred in Jatitujuh.

PSGM, Hybridized in Pasuruan, East Java and Indonesia Sugar Research Center bred in Gunung Madu.

PSTK, Hybridized in Pasuruan, East Java, and Indonesia Sugar Research Center bred in Takalar.

PSR, Philippines.

Q, Hybridized in Queensland Meringa sugar experimental station, BSES bred.

R, Crossed in Coimbatore, India, bred in Rudrur (AP).

    R (Rarawai), Fiji Sugar Company.

    R (Reunion), Reunion Island, Africa.

    RA, EEAOC-INTA , Argentina.

    RB, Federal University of Argus, Brazil.

    RBB, Bolivia.

    RD, Dominica.

    RK, Japan.

S, Saipan Island, Japan.

    S, Hybrid in Coimbatore, India, and varieties cultivated in Shahjahanpur (UP).

    SA, South Africa.

    SC, Saint Croix, Virgin Islands.

    Si, Hybrid in Coimbatore, Indian, and varieties cultivated in Sirugamani (TN).

    SJ, South Johnston, Queensland.

    SK, St.Kitts.

    SL, Sri Lanka Sugarcane Institute.

    SP, Sao Paulos.

    SPF, Sugarcane Research Institute in Pakistan.

    SPSG, Shakarganj Sugar Research Institute in Pakistan.

    SW, SempelWadak, Java

T, Hybrid in Coimbatore, Indian, and varieties cultivated in Tirupati (AP).

    T, Trinidad.

    Tiep, Tjepiring, Java.

    TC, Malaysia.

    TCP, Texas A&M University.

    THAI, Thailand (SFCRC).

    Tuc, Tucuman, Argentina.

UD, Hybrid from the Hawaiian Yuba (UBA) series (UBA × D1135).

UM, Varieties and materials were collected or developed by the University of Malaysia (UM).

US (U.S.Experiment Station), US Department of Agriculture Experimental Station.

UT, Varieties bred by UT Field Crop Research Center, Suphan, Thailand.

V, Fonaiap, Venezuela.

VMC, Victoria Milling Company.

Wi, Hybrid in Barbados, varieties bred in Barbados and the West Indies.

ZN, Sugar Association in Zimbabwe.

6.3 Introduction on New Sugarcane Varieties

Sugarcane is the primary raw material of the sugar industry. Sugarcane varieties for sugar production should be high-yield, high-sugar content, good ratooning, strong stress resistance, and have good agronomic and technological characteristics. The improvement of sugarcane variety has also continued to develop with the rise and development of the sugar industry in China. There are more than 240 sugarcane varieties through the introduction and sexual hybrid breeding since the founding of new China 60 years ago, including different maturity periods and suitable for various ecological types, from sugarcane scientific research institutions in main sugarcane-producing provinces (regions). These varieties have evolved from landraces to introduced varieties, self-breeding varieties, and the promotion and renewal of new varieties. They play an essential role in promoting the sustainable development of the sugar industry.

According to the variety's maturity period, some new sugarcane varieties have been bred recently, and some promotion areas are classified and introduced.

6.3.1 Early maturing varieties

6.3.1.1 'CYZ 05-51'

Combination: 'CYC 90-56' × 'ROC23'.

Breeding organization: Sugarcane Research Institute, Yunnan Academy of

Agricultural Sciences, and Yunnan YunZhe Technology Development Co., Ltd. It passed the national sugarcane variety identification in 2013.

Features: The plant is tall and upright. The stem is medium or large, and the stem content is solid. The internode length is medium, cylindrical. The wax powder of the stem surface is thick. There are no water cracks and air roots on the stem surface. The stem color is yellow-green before internode exposure and purple after exposure. The bud body is rhombus and medium. The bud groove is shallow and not obvious. The bud wing is medium, and the bud tip is overtopped the growth zone. The bud base is flat with a leaf scar. The root zone is moderate. The leaf tip is drooping. There are fewer or no No.57 hair groups. The inner ear is triangular, and the outer leaf ear is missing.

Characteristics: This variety is early-maturing and high-sugar with fast emergence, strong tillering. The sugarcane stalks are uniform and neat. It is ratoonability firm defoliation, high resistance to smut, medium resistance to mosaic, and strong drought resistance. In the national, regional test, the average sugarcane yield was 100.77 t/hm^2, and the average sugar yield was 15.18 t/hm^2. The average sugar content was 14.10% from November to December, 15.67% from January to March, and 15.00% in the whole period. In the production test, the average sugarcane yield was 102.60 t/hm^2, and the sugar yield in the whole period was 15.02 t/hm^2. The average sugar content was 13.88% from November to December, 15.24% from January to March, and 14.68% in the whole period.

Key points of cultivation: The suitable row spacing is 1.1−1.2 m. The number of planting buds should be 56,000−68,000 double buds per hectare; however, appropriately would be increased in the dry land. In dry slope land, deep furrow planting and subsoil cultivation should be adopted. Film mulching should be used for winter or early spring planting; The variety should be pursued through fertilizer in the early seedling stage. It should be applied foot-tapping fertilizer in the moderate growth stage and be suitable for high soil cultivation to prevent late lodging. The management of ratoon ability should be strengthened. The sugarcane field should be cleaned timely after the harvest, irrigated, and loosen stumps early, while film mulching promotes early-onset, multiple seedlings. To make full use of soil moisture and promote the germination of sugarcane plants, sugarcane leaves or plastic film should not be used to cover the stump after harvesting in dry slope land. Attention should be paid to control dead heart seedlings at the seedling stage and thrips at the large growth stage.

6.3.1.2 'CYZ 08-1609'

Combination: 'CYZ94-343' × 'CYT00-236'.

Breeding organization: Sugarcane Research Institute, Yunnan Academy of Agricultural Sciences, and Yunnan YunZhe Technology Development Co., Ltd.

Features: This variety is early maturity, high yield, and high sugar content. The stem is medium and large. Plant type is compact, erect. Internode is conical. Cane stem yellow-green, yellow-green after exposure, bud shape pentagonal, bud tip reaching growth zone, bud wings arc-shaped, bud base is flat with the leaf mark, Leaf arch, leaf width, green, sheath anthocyanin color intensity, less hair, inner ear lanceolate, outer ear triangle, stem solid, good defoliation.

Characteristics: The seedling emergence is neat and strong, the stem rate is high, the sugarcane plants are uniform and neat, the plant type is compact, the leaves are clear, the drought resistance is good, and the ratoonability is strong. The yield of sugarcane was more than 105.0 t/ hm^2. In November, the average sugar content was 14.12% at multiple sites for many years. In the mature period (from November to March), the average sugar content was 16.20%, 1.36% higher than that of the control (14.84%) (absolute value). In March 2016, the sugar content reached 21.02%, and the sugar yield reached 18.0 t/hm^2 in the Gengma Base of the Lincang Sugarcane Test Station. This variety's sugar content and gravity purity were higher than 'ROC22' within 1-7 days after harvest at the maturity stage, and the transformation resistance was excellent. It is suitable for planting in dry slope land, platform dam land, and paddy with better water and fertilizer conditions. It is recommended not to plant in the high incidence area of smut.

6.3.1.3 'CLC 05-136' ('Guiliu05136')

Combination: 'CP81-1254' × 'ROC22'.

Breeding organization: It was bred by Liucheng Sugarcane Research Center and passed the national sugarcane variety identification in April 2014.

Features: This plant is height, compact moderate, large stems, cane stem erect uniform. The bud body is medium, round, lower born in leaf mark, bud tip to growth zone, bud wing lower edge to bud 1/2, bud hole born in bud body upper-middle, root 2 rows. Roots are purple-red, growth yellow-green. Internode is cylindrical. The stem color is yellow-green on the shaded part, purple on the exposed part. The stem surface has more powder, and the bud groove is shallow. Stems are solid with shallow growth cracks (water cracks). Leaves are erect, green, and their sheath is purple-red. There are more No.57 hairy groups. Outer auricles are transitional, and inner auricles are triangular. The leave sheath is easy to defoliate.

Characteristics: Two years of new planting and one year of ratooning test in the national sugarcane variety regional test from 2012 to 2013, the average yield of sugarcane was 100.74 t/hm^2, which was 0.87% higher than that of the control variety 'ROC22', in which was increased by 0.15 % in the new planting season and increased by 2.1% in the ratoonability season. The average sugar yield was 15.16 t/hm^2, which was 5.41% higher than 'ROC22'. The average sucrose content of sugarcane from November to March next year was 14.99%, 0.57 percentage points higher than CK. The average sucrose content of sugarcane from November to March next year was 15.01%, 0.53 percentage points higher than CK. The average yield in 2013 was 94.93 t/hm^2, which was 2.75% lower than that of 'ROC22' (CK).

Key points of cultivation: the best row spacing is 100−120 cm. It is suitable to plant 45,000 segments of double buds per hectare. Soaking seeds with 0.2% carbendazim solution for 3−5 minutes is best to prevent pineapple disease. At the same time, we should apply pesticides to control underground pests and cover the film after covering the soil to achieve heat and moisture preservation. Attention should be paid to the rational application of nitrogen, phosphorus, and potassium to avoid partial and heavy nitrogen fertilizer application and ensure adequate phosphorus fertilizer. Fertilization time as early as possible, new sugarcane should be in late May fertilization, ratoon sugarcane in mid-April fertilization. Early pest control and early fertilization management should retain more than three years of ratoonability, improving sugarcane planting efficiency. Field management should be timely and in place. The variety grows fast at the jointing stage. Attention should be paid to the control of sugarcane borer.

6.3.1.4 'CGT42' ('CGT04-1001')

Combination: 'ROC22' × 'CGT92-66'.

Breeding organization: It was bred by the Sugarcane Research Institute of Guangxi Academy of Agricultural Sciences and passed the Guangxi sugarcane variety certification in 2013.

Features: The plant is erect, uniform and the stem are medium and large. The stem color is light yellow on the shading part and purple-red on the exposure part. Solid, cylindrical internode with medium length, rhomboid bud, bud groove not obvious, bud top flat or exceeding growth zone, bud base falling into leaf mark, bud wing large, leaf angle is small, leaf blade green, sheath length medium, leaf sheaths fall off easily, inner auricle triangle, no outer ear, No.57 hairs group short, few or none.

Characteristics: Early and middle maturing varieties, good germination and emergence, early growth and rapid development, high tillering rate, more effective stems, good defoliation. From 2011 to 2012, they participated in Guangxi regional test, two years of new planting and one year of ratoonability test, and the average sugarcane yield was 101.7 t/hm^2, which was 9.26 % higher than that of the control 'ROC22'; The sugar yield was 150.3 t/hm^2, which was 14.45% higher than that of 'ROC22'. The average sucrose content of sugarcane from November to February next year was 14.77%, 0.66% higher than the control (absolute value). High and stable yield, good ratoonability, wide adaptability, lodging resistance, drought resistance, high resistance to shoot rot.

Key points of cultivation: medium and large stem species, planting row spacing of 1.0−1.2 meters, 120 thousand buds per hectare; Application of base fertilizer, early topdressing, to meet the needs of the early growth of varieties while strengthening the management of middle tillage, paddy field appropriate high soiling to prevent late lodging; To effectively utilize the deep soil moisture, deep trench planting should be adopted in dry slope land cultivation. Winter planting or early spring planting needs film mulching cultivation. Strengthen the management of ratoon crop, cane harvest before the season should be timely cleanup cane fields. Irrigation conditions cane orchard should be done early irrigation, early loosening of stumps, film mulching to promote the germination of ratoon crop, to ensure ratoon crop high yield; Sugarcane in dry slope land was cut with the sharp hoe, and then covered with sugarcane leaves or film to make full use of soil moisture and promote sugarcane plant germination.

6.3.1.5 'CYZ 99-91'

Combination: 'ROC10' × 'CYC 84-125'.

Breeding organization: Sugarcane Research Institute, Yunnan Academy of Agricultural Sciences, and passed the national sugarcane variety approval in 2010.

Features: plant erect, medium and large stems, plant medium-high, stalk up and down uniform, internodal cylinder slightly thin waist, internodal length, internodal exposure before and after yellow-green, no stripes and growth cracks, root zone width medium, green color, growth with gray orange, bud top shape is convex, bud shape is round, bud body is large, no bud groove, no bud tip 10 hair group, bud tip from growth zone, bud base connecting leaf scar, leaf sheath length medium, No. 57 hairy more; The leaf tongue was strip-shaped, and the inner and outer auricles were transitional, with brown rectangular thickening zone, leaf blade length, and width medium, leaf margin serrated, leaf straight, etc.

Characteristics: early maturing high sugar varieties, high emergence, general tillering ability, moderate plant height, large diameter. The average sugarcane yield was 100.65 t/hm^2, and the average sugar content was 15.80%. The sugar yield was 15.81 t/hm^2. This variety has a high and stable yield. It is suitable for planting in sugarcane fields with medium fertility and good water and fertilizer conditions in Yunnan, Guangxi, Guangdong, and Fujian.

Key points of cultivation: It is suitable for planting in terrace, irrigated land, dam land, paddy field, and deep dry slope land with good water and fertilizer conditions, and the number of double bud segments per hectare is 60,000–68,000; Apply enough base fertilizer, early topdressing, strengthen fertilizer and water management to promote seedling strong; Buds big, keep seedlings, prevent mechanical damage. Sugarcane should be sharp hoe low cutting, early fertilization management, promote early growth and rapid development.

6.3.1.6 'CYT93-159'

Combination: 'CYN73-204' × 'CP72-1210'.

Breeding organization: It was bred by Zhanjiang Sugarcane Experiment Station, Sugarcane Industry Research Institute, China Light Industry Association, and passed the national sugarcane variety certification in 2002.

Features: Medium to large stems, solid, stem diameter uniform, stem shape beautiful, base thicker, internodes longer, cylindrical, cane stem not exposed part of the blue-yellow, yellow-green after exposure; Sugarcane stem without water crack, no air root, no growth crack, and bolt plaque, bud oval, a base near leaf scar, top does not reach the growth zone; Leaf blade green, medium width, new leaf erect, old leaf bowed, easy to defoliate, leaf sheath green or slightly yellowish-green. Inner ear lanceolate, outer ear triangular.

Characteristics: Early maturing high sugar content, high purity varieties, early germination fast and neat, high germination rate, strong tillering ability. Fast growth, early ridge closure, superior growth, good hairy, strong ratoonability, high resistance to smut, mosaic disease. The average sugarcane and sugar yields were 89.5 t/hm^2 and 14.9 t/hm^2, respectively, 17.3% and 30.7% higher than 'ROC10'. In December, the sugar content of sugarcane was 15.86% in November, 16.78%, and 17.43% in January, respectively, 1.96%, 1.51%, and 1.68% higher than that of 'ROC10' (absolute value). From November to January, the following year averaged 16.69%, 1.72% (absolute value, the same below), than 'ROC10'.

Key points of cultivation: this variety is suitable for cultivation in paddy fields or irrigated land with medium or above soil fertility and suitable temperature and heat conditions; It is appropriate to control the number of effective stems at 75,000 for 60,000 double-bud segments per hectare; Suitable for winter planting or early planting, cover film when planting; Sugarcane plant early and strong ratoonability, should retain more than 2 years of perennial, to improve the efficiency of sugarcane planting.

6.3.1.7 'CYT00-236'

Combination: 'CYN73-204' × 'CP72-1210'.

Breeding organization: It was bred by Zhanjiang Sugarcane Research Center of Guangzhou Sugarcane Research Institute and passed the identification of the National Variety Approval Committee in 2010.

Features: Medium to medium stem, solid, base coarse, internodes slightly conical, no bud groove. The unexposed part of sugarcane is light yellow, and the exposed part is blue and yellow after sunshine exposure. The stem surface is covered with thin white wax powder: stem diameter uniform, no air root, no growth cracks, and bolt plaque. Buds smaller, ovate, base leaflet scarred, apex not reaching growth zone. Bud wing width medium, born in the upper part of the bud, bud hole near the top. Leaves pale green, leaves slightly narrow, slightly short, heart leaves erect, old leaves scattered. Leaf-sheaths green, hairy group 57 underdeveloped. Inner ear lanceolate, outer ear triangular.

Characteristics: Early maturing and high sugar content varieties. Germination fast, neat, tillering solid ability, high stem rate, the number of raw stems, the whole growth period of robust growth, late not premature aging, strong ratoonability. The average sugarcane yield was 105.03 t/hm^2, and the sugar yield was 16.30 t/hm^2. The average sucrose content was 15.52 % in January next November, and the average sucrose content in the whole period was 16.05 %. Resistance to smut, high resistance to mosaic disease.

Key points of cultivation: The variety is suitable for paddy fields and irrigated land with medium or above soil fertility and suitable for planting in winter or early spring. This variety has a high germination rate and tillering solid ability. It is suitable for sparse planting and 40,000−45,000 double bud seedlings per hectare. Pesticides should be applied once for new sugarcane planting and medium soil culture; Sugarcane plant early and more should be early pest control, early fertilization management; Nitrogen, phosphorus, and potassium should be applied together, and phosphorus fertilizer should be appropriately increased to avoid partial application and heavy application of nitrogen fertilizer.

6.3.1.8 'CLC03-182'

Combination: 'CP72-1210' × 'ROC22'.

Breeding organization: Guangxi Liucheng Sugarcane Research Center and passed the national sugarcane variety identification in 2011.

Features: Plants erect, long internodes, compact and moderate plant type, strong vigor, not premature senescence. Buds medium, tender stem yellow-white, exposure, yellow-green, old stem dark yellow, wax powder thick, not hollow or pith. The leaf shape is narrow, leaf thickness, leaf color is dark green, leaf posture upright, leaf upper slightly drooping, leaf canopy distribution space is not easy to shade each other, good transparency. No.57 hairs are so many, but leaves fall off after maturity, easy to defoliate.

Characteristics: Early maturing varieties, early germination, early jointing, strong vigor, medium tillering ability, high stem rate, can make full use of early light, temperature, and water to obtain a high yield, high sugar content. Late October can reach the mature technology standards, sugar content decreased slowly in the late crushing season. The average sugarcane yield was 97.5 t/hm^2, and the sugar yield was 14.8 t/hm^2. The sugar yield reached 14.5% in November, 15.83% from November to January next year, and the peak period was above 18.0 %. The frost resistance was strong, and the drought resistance was equivalent to 'ROC22'. It is sensitive to smut. This variety is sensitive to diuron and atrazine, easy to produce phytotoxicity, significantly delayed growth.

Key points of cultivation: Suitable for the general level of sugarcane planting, the water and fertilizer conditions of medium or higher sugarcane field can play its effect of increasing yield and sugar; Shoot head was selected for cultivation as far as possible, and 45,000−53,000 double bud seedlings were planted per hectare. Root hair early should be early management, early fertilization. It was very sensitive to diuron. When herbicides such as ametrine were applied at the seedling stage, spraying to sugarcane plants should be avoided, and the recommended dosage should be strictly applied.

6.3.1.9 'CCZ19' ('CCT 79-15')

Combination: 'CCT 61-380' × 'CCZ 4'.

Breeding organization: It was bred by Sichuan Sugar Manufacturing and Sugar Industry Research and passed Sichuan sugarcane variety certification in 1993.

Features: plants erect; Sugarcane medium, medium stem species; Internode cylindrical, no growth cracks and bolt stripes; Leaf-sheath wrapped part of yellow-green, light part of green; The wax powder exposed part was thin, and the old leaf sheath wrapped part was mostly black; Buds round; Roots green; Leaf blade mid-wide smooth, mid-upper bent down; Leaves sheath smooth without 57 hairs, easy to fall off, conducive to harvest; inner ear lanceolate, outer ear triangular; 5−7 leaves thick yellow-green, square; Leaf posture oblique scattered.

Characteristics: Early maturing, high sugar content varieties. Sugarcane buds germinated fast and neatly; Tillering capacity slightly weak; the growth rate was fast in the early and middle stages and decreased in the late stage; Strong cold resistance; strong ratoonability; Strong fertility resistance; Good group uniformity; Resistance to smut and mosaic disease, resistance to brown stripe disease, more resistant to moth. There are more effective stems, with an average of 75,000–90,000/hm^2, sugarcane yield 76–90 t/hm^2, up to 113.3 t/hm^2; Sugar content is higher; generally 13.63%–14.14%, up to 14.43%; The average sugar yield was 10.5–12.0 t/hm^2.

Key points of cultivation: The variety is suitable for planting in the hilly dry land, river dam land, terrace, and other lands in the central subtropical sugarcane area, and it can better play its high yield performance in the groove dam land with high fertility and good water and fertilizer conditions. Cultivation of plastic film mulching, plantar fertilizer, early topdressing, tillering fertilizer, shoot fertilizer, do a good job of soil.

6.3.1.10 'ROC16'

Combination: 'F171' × '74-575'.

Breeding organization: Taiwan Sugarcane Industry Research Institute. It was approved by the National Variety Approval Committee in 2002 and was approved by the Yunnan Variety Approval Committee in the same year.

Features: Medium to large stem species. Internodal cylindrical, stem color before exposure yellow-green, after the initial light purple, late light yellow. Wax powder with thick, no cracks and bolt spots, bud groove shallow, growth belt slightly prominent; Bud egg round, bud base sink into leaf mark, bud top flat growth zone, bud wing narrow; Leaves erect at the seedling stage, tip bent, young leaf sheath pale purple, old leaves turn gray-green, inner ear short lanceolate, outer ear flat transition type, easy to defoliate, leaf sheath back hairy.

Characteristics: Early maturity, high yield, and high sugar content varieties. Germination rate is high, effective stem more, stem uniform, no pith, not easy to lodging and heading. The yield of sugarcane was above 90 t/hm^2. The sucrose content of sugarcane reached about 13.5% in November, 14.8% on average from November to January next year, 15.6% in the peak period, and no sugar was returned in the later period. Strong ratoonability, whether autumn planting, perennial sugarcane yield, is stable. Resist mosaic disease, smut, susceptible to brown spot in hot and humid season.

Key points of cultivation: Suitable for medium or above fertility soil cultivation, irrigation conditions cane area is better; Strong tillering ability, suitable sparse planting and delay the operation time of soil culture, so that the late tillering total growth; Appropriate control of nitrogen application and exclusion of field waterlogging.

6.3.1.11 'ROC20'

Combination: '69-463' × '68-2599'.

Breeding organization: Taiwan Sugarcane Industry Research Institute, Approved by the Yunnan Variety Approval Committee in 2002.

Features: Medium to small stem variety, internodes cylindrical, slightly curved near internodes, pale purple stem color, purple-red to dark purple after exposure, no growth cracks and wood embolus patches in stem bark, no prominent bud groove, and light yellow growth zone slight protrusions; Buds small, elliptic, bud base on leaf scars, apex flat growth zone, bud wings small, thin and narrow; Root zone light yellow to dark purple, root point two rows arranged neatly, occasionally the third row; Leaves verdant green, leaf width medium, young leaf sheath light blue-purple, 57 hairs obvious, late fall off, easy to defoliate, not easy to heading, not easy to lodging.

Characteristics: Early maturing, high sugar content varieties. Germination is rapid, neat, tillering, full growth vigor stems and leaves lush, more effective stems, cane stem erect. According to Taiwan, the sugar content can reach 13 % in October and more than 15 % in December. The average sucrose content from October to April next year is 14.74 %, 0.9 %, 1.25 %, and 0.99 % higher than 'ROC1' 'ROC10' and 'ROC25', respectively. The yield is 10 %–17 % and 6 %–14 % higher than 'ROC1' and 'ROC10', respectively. It is suitable for spring planting, autumn planting, and ratoonablity cultivation and has good ratoonablity.

Key points of cultivation: In Taiwan, to adapt to the western coastal plain and red soil platform, irrigation and drainage of good sandy loam, loam, and clay loam planting; The introduction time was not long, and the research and experimental demonstration of variety characteristics, adaptation and supporting cultivation techniques should be done well. This variety is susceptible to borers, and special attention should be paid to the prevention and control of borers at the seedling stage.

6.3.1.12 'ROC25'

Combination: 'PT79-6048' × 'PT69-463'.

Breeding organization: Taiwan Sugarcane Industry Research Institute.

Features: The stalk is long, and the internode is cylindrical. The sugarcane stem is light yellow before the leaf sheath abscission, and the leaf sheath is light purple after long sunshine exposure. Stem surface covered with white wax powder, buds medium size. The young buds before leaf sheath abscission were light yellow oval. The old buds protruding outwards after leaf sheath abscission were dark purple. The bud base followed the leaf mark, and the growth zone was the horizontal line at the top of the bud. It passed through the back of the top of the bud, and the width of the bud wing was medium. It was born in the upper half of the bud, and the bud hole was located in the upper half, with No.7, No.8, No.10, and No.16. Leaf-blade dark green, leaf width medium, leaf posture erect, leaf oblique, new leaf tip erect, old leaf tip bent. Not easy to fall off, leaf sheath green, old leaf sheath slightly purple, leaf sheath covering wax powder unknown, 57 hairs underdeveloped.

Characteristics: Early-maturing varieties. Germination neat, tillering robust and early growth fast, growth advantage, raw stem and long, stem diameter medium, easy ridge closure, easy to control weeds, strong ratoonability. The average sugarcane yield was 108.3 t/hm^2, and the sugar yield was 13.85 t/hm^2. The average sugar content was 14.40% in November and 15.13 %–15.53 % from December to February. In the ratoon experiment, the sugarcane and sugar yield increased by 14% and 30% on average compared with the control 'ROC10'. Strong resistance to leaf blight, low resistance to yellow rust.

Key points of cultivation: It has good growth potential, easy to seal ridge, easy to control weeds, strong ratooning power, and easy to manage; The number of aphids was more than that of 'ROC10', so special attention should be paid to control; Suitable for medium or above fertility soil cultivation.

6.3.2 Mid-maturing varieties

6.3.2.1 'CYZ 03-194'

Combination: 'ROC25' × 'CYT97-20'.

Breeding organization: Sugarcane Research Institute, Yunnan Academy of Agricultural Sciences, and Key Laboratory of Sugarcane Genetic Improvement in Yunnan Province and passed the national sugarcane variety identification in 2011.

Features: The plant is tall and erect, with medium and large stem, long internode, cylindrical, gray, no water crack, air root, no hollow, no cattail, no No. 57 hair group; Bud rhombic, bud body large, bud groove shallow not obvious, bud tip to the growth zone, bud base is flat with the leaf scar; Root point 3 rows, more regular; Leaves scattered, leaves thick green, long, wide; The inner ear is triangular, and the outer ear is absent.

Characteristics: Early-middle maturing high-yield and high-sugar varieties, fast emergence, neat, uniform, good tillering, a strong ratoonability, good defoliation, high resistance to smut, strong drought resistance. No major sugarcane diseases such as mosaic disease and yellow leaf syndrome were found under natural conditions in the field. The average sugarcane yield was 106.21 t/hm^2, the sugar yield was 14.89 t/hm^2, and the sugarcane sucrose content was 13.82 % from November to December and 15.12 % from January to March, and the whole period average was 14.47%.

Key points of cultivation: Planting row spacing was 1.1–1.2 m, and 120,000 buds per hectare. For winter or early spring planting, it is necessary to cover with film, apply sufficient base fertilizer, top dressing at the seedling stage, apply sufficient stem attacking fertilizer at the moderate growth stage, and adequately cultivate high soil to prevent lodging late stage. Strengthen the management of ratoon crops, sugarcane harvest before the season should be timely clean up sugarcane field, irrigation conditions to achieve early irrigation, early loosening stumps; Cane stumps should not be covered directly with cane leaves or plastic film after harvest in dry slope land to make full use of soil moisture to promote sugarcane plant germination.

6.3.2.2 'ROC10'

Combination: 'ROC5' × 'F152'.

Breeding organization: Taiwan Sugarcane Industry Research Institute was introduced into mainland provinces (autonomous regions) successively in the mid-1980s.

Features: Medium and large stems, plant erect and compact. Stems shade part pale green, exposure part gray-green. Internode conical, micro zigzag. Wax powder belt obvious, growth ring medium prominent. Buds are pentagonal, full and prominent. Bud tip reaches growth zone. Bud base leaves a mark, bud ditch is obvious, leaf ribs developed, vertical not drooping, dark green. No auricles, no No.57 hairy group.

Characteristics: It belongs to mid-early maturity, high yield, high sugar, and multi-resistance variety. Germination is fast and neat; early growth is slow, the stems are uniform, stem length is medium, stem number is many, not hollow, not bush heart, not ending, a strong ratoonability, high resistance to smut, downy mildew and rust, insect resistance, wind resistance, not easy to fall. The yield of 'ROC10' was 97.5–117 t/hm^2, which was more than 10% higher than that of field variety 'Xuan 3' and 'CMT 70-611'. The average sugar content of 'ROC10' was 14.63%, which was more than 0.58 % higher than that of CK.

Key points of cultivation: This variety is suitable for planting in sugarcane fields with high temperature and medium water and fertilizer conditions. It is suitable for autumn or winter planting plastic film mulching cultivation, avoid late spring planting; Appropriate close planting, stem control in 90,000 plants/hm^2 is appropriate; In the middle and late stages, nitrogen fertilizer application and irrigation times should be appropriately controlled to promote sucrose accumulation, an appropriate extension of perennial life to improve economic efficiency.

6.3.2.3 'ROC22'

Combination: 'ROC5' × '69-463'

Breeding organization: Taiwan Sugarcane Industry Research Institute. It passed the national sugarcane variety approval in 2002 and the Yunnan provincial variety approval in the same year.

Features: Medium to large stems, base coarse, tip head small, stem long, internode inverted conical; Sugarcane showed light yellow-green before leaf stripping, purple-red at the beginning of leaf stripping, dark purple-red after sunshine exposure; Wax powder belt with thick wax powder, the inter-node wax powder is also thick, evenly distributed; Sugarcane has no growth crack and no bolt patch; Buds ovate round, shoot apex overgrowth, bud wings medium-wide; The bud groove is very obvious, from the top of the sugarcane bud to the leaf mark; Growth zone slightly protruding, pale yellow; Root 2–3 column, irregular arrangement. Leaf-sheaths cyan purple, inner ear lanceolate, outer ear obtuse triangular.

Characteristics: Medium maturity, high sugar content, good germination, high tillering rate, slow growth in the early, fast growth in the late. No. 57 hairs developed, easy to defoliate, no easy to lodge, no heading flowering, strong ratoonability. Resistant to smut, leaf blight, brown rust, moderate resistance to mosaic disease, and moderate response to sugarcane aphid. The average sugarcane yield and sugar yield per hectare were 103.5 t/hm^2 and 14.5 t/hm^2, and the sugar content was 14.03 %.

Key points of cultivation: The number of effective stems was controlled at 82,000 – 90,000 with the seed amount of 45,000–50,000 double bud segments per hectare;

Can be planted in advance, using seedling disinfection, germination, film mulching cultivation techniques; The ratoonability is strong. It should be opened and managed early, and the cultivation period of the ratoon crop can be appropriately prolonged. Pay attention to control thrips and pokkah boeng disease.

6.3.3 Late-maturity varieties

6.3.3.1 'CYZ 06-407'

Combination: 'CYT97-20' × 'ROC25'.

Breeding organization: Sugarcane Research Institute, Yunnan Academy of Agricultural Sciences, and Yunnan YunZhe Technology Development Co., Ltd. It passed the national sugarcane variety identification in 2013.

Features: Plant tall and erect, medium and large stem, solid, internode length medium, cylindrical, wax powder thick, no water crack, no air root, internode exposure before the yellow-green, after exposure purple-red; Bud egg round, bud body medium, bud groove not prominent, bud wings small, bud tip slightly reaching growth zone, bud base is flat with the leaf mark; Root point 3 rows, more regular; Leaf blade thick green, leaf posture oblique set, leaf tip drooping, No. 57 hairs more, easy to fall off; Inner ear lanceolate, outer ear absent.

Characteristics: Late maturing varieties, quick emergence, neat, uniform. The average sugarcane yield was 110.51 t/hm^2 in the 8th National Regional Trial from 2011 to 2012, which ranked first, increased by 17.45% compared with 'ROC16' and 10.85 % 'ROC22'. The sugar yield was 15.28 t/hm^2, 3.38 %–11.28% higher than two CKs. The sugar content was 13.01% from November to December, 14.54 % from January to March, and the peak was above 16.5%. Strong ratoonability, many effective stems, good lodging resistance, good defoliation, high resistance to smut disease, sugarcane streak mosaic virus (S_{cs}MV) showed high sensitivity, sorghum mosaic virus (SrMV) showed immune.

Key points of cultivation: Planting row spacing 1.1–1.2 m, 60,000 double bud segments per hectare; The dry land is cultivated by deep furrow planting, winter planting or early spring planting need plastic film mulching cultivation; Application of organic fertilizer or filter mud as base fertilizer, early topdressing at the seedling stage, applying enough fertilizer to promote sugarcane growth in the moderate growth stage; Strengthen the management of ratoon crop, harvest before the season, should be timely cleanup sugarcane field, irrigation conditions to achieve early irrigation, early loosening, film mulching to promote early-onset, multiple plants. Cane stumps should not be covered directly with sugarcane leaves or plastic film after harvest in dry slope land to use soil moisture and promote sugarcane germination fully.

6.3.3.2 'CYT 86-368'

Combination: 'F160' × 'CYT71-210'.

Breeding organization: It was bred by Zhanjiang Sugarcane Experimental Station of Sugarcane Sugar Research Institute of General Council of Light Industry. It was approved by the National Variety Approval Committee in 2002.

Features: Medium and large stem solid, plant tall and upright, cane stem smooth without water crack, no air root, stem shape uniform and beautiful. Internodes long, cylindrical, young part purple, brown purple after exposure, wax powder thick, narrow root zone, small root point, 2-3 rows arranged irregularly. Buds are oval or nearly round, not easy to aging, bud base near leaf marks. The top does not exceed the growth zone, no bud groove. Leaves green, slightly short, scattered, leaf sheath green and slightly purple, sheath back glabrous, old leaves easy to fall off, hypertrophy with larger triangular, inner ear short hook, outer ear transitional.

Characteristics: Mid-late maturing and high sugar varieties. It has good germination and tillering. The seedlings are neat and robust; jointing growth is rapid, vigorous growth, more stems, thick and uniform, about 75,000 stems per hectare, the average cane yield is about 120 t/hm^2. In November, December, and January, the sugar content of sugarcane is 11.24%, 14.73%, 15.25%, and 14.14%. The root system is developed, drought-resistant, and the growth advantage is obvious in the dry season, not easy to windbreak and lodging. It has strong resistance to mosaic and smut.

Key points of cultivation: Strong drought tolerance and suitable for planting in various sugarcane areas, especially in dryland with medium or higher fertility. The variety is tall and vigorous after growth. It is suitable to plant early, strengthen the management of topdressing at the seedling stage, and promote root system development to lay a good foundation for later growth. Sugarcane fields with seedling roots should not be harvested on rainy days or when the soil is too wet; otherwise, the hairy roots are vulnerable.

6.3.3.3 'CFN 91-4621'

Combination: 'CP72-1210' × 'CZZ 74-141'.

Breeding organization: Sugarcane Comprehensive Research Institute of Fujian Agriculture and Forestry University and passed the national sugarcane variety approval in 2002.

Features: Medium-large stalk, solid, plant erect, internode cylindrical, no water crack, yellowish-brown shading stem skin part, purple-red after exposure, no air rooting and growth cracks, wax powder slightly more. Buds round, bud base leaves a mark, bud tip and growth band flat, bud wings more minor, no bud groove. Growth zone slightly prominent, green, turned purple after exposure, leaves green, leaf width medium, yellow-green leaf sheath, no No. 57 hairs group, base slightly purple-red. Leaves developed, inner and outer ear lanceolate, inner ear longer.

Characteristics: Late maturing varieties, germination early and neat, germination rate is high, seedling strong, tillering medium, stem base thick, lodging resistance is strong, the early emergence of ratoon crop, suitable number of seedlings, substantial ratooning. Cane sugar 14.3%. Inoculation identification, race 2 with high resistance to smut, race one susceptible to smut, high resistant mosaic disease. The average sugarcane yield was 103.1 t/hm^2, the average sugar content was 13.56%, and the sugar yield was 13.97 t/hm^2.

Key points of cultivation: Fields with water irrigation above medium-fertility were selected for planting, and 45,000 double-bud segments were planted per hectare. Application of enough basal fertilizer, early topdressing, and tillering promotes tillering into the stem and appropriately delays the fertilizer stop period.

References

LI M Z, HUANG J K, 1999. Breeding of a new sugarcane variety Chuanzhe19. Sugar Crops of China (3):6-8.

LI Q W, LIU F Y, CHEN Y G, et al., 2012. Study on Breeding and Characteristics Analysis of New Sugarcane Variety Yuetang 03-393 (Yuetang No. 60). Southwest China Journal of Agricultural Sciences, 25 (2):401-407.

LI Y R, 2010. Modern Sugarcane Science. Beijing:China Agriculture Press (in Chinese).

PENG S G, 1990. Sugarcane Breeding. Beijing:Agriculture Press (in Chinese).

WU C W, ZHANG Y B, 2009. New technology for high-yield cultivation and processing of sugarcane. Kunming:Yunnan Science and Technology Press (in Chinese).

WU C W, ZHAO P F, XIA H M, et al., 2014. Modern cross breeding and selection techniques in sugarcane. BeiJing:Science Press (in Chinese).

WU W L, LIU F Y, PAN F Y, et al., 2011. Breeding of YT03-373, an early-mid new sugarcane variety. Guangdong Agricultural Sciences (21):41-43.

ZHANG Y B, GUO J W, HUANG Y K, et al., 2010. High-yield sugarcane production technology. Kunming:Yunnan Science and Technology Press (in Chinese).

ZHANG Y B, WU Z K, LIU S C, 2004. Standardized Comprehensive Cultivation Techniques of High Sugar and High Yield Sugarcane in Yunnan Province. Kunming:Yunnan Science and Technology Press (in Chinese).